The Craft of Scientific Writing

Third Edition

The Craft of Scientific Writing

Third Edition

Michael Alley

Includes 28 Illustrations

Michael Alley
Mechanical Engineering
Electrical and Computer Engineering
Virginia Tech
Blacksburg, VA 24061-0238
alley@vt.edu

Cover illustration: Parachute system designed for the crew escape module of the F-111 fighter jet. Peterson and Johnson, 1987. This illustration also appears on p. 162 of the text.

Ancillary information for the book can be found through the publisher's website: http://www.springer.ny.com

Library of Congress Cataloging-in-Publication Data
Alley, Michael.
The craft of scientific writing/Michael Alley. — 3rd ed.
p. cm.
Includes bibliographical references and index.
ISBN 0-387-94766-3 (soft : alk. paper)
1. Technical writing. I. Title
T11.A37 1996
808'.0666 — dc20 96-15207

ISBN-10: 0-387-94766-3
ISBN-13: 978-0387-94766-2

Printed on acid-free paper.

First edition published by Prentice-Hall, Inc., 1987. Second edition published by Michael Alley, 1995.

Printed in the United States of America. (DH)

9

springeronline.com

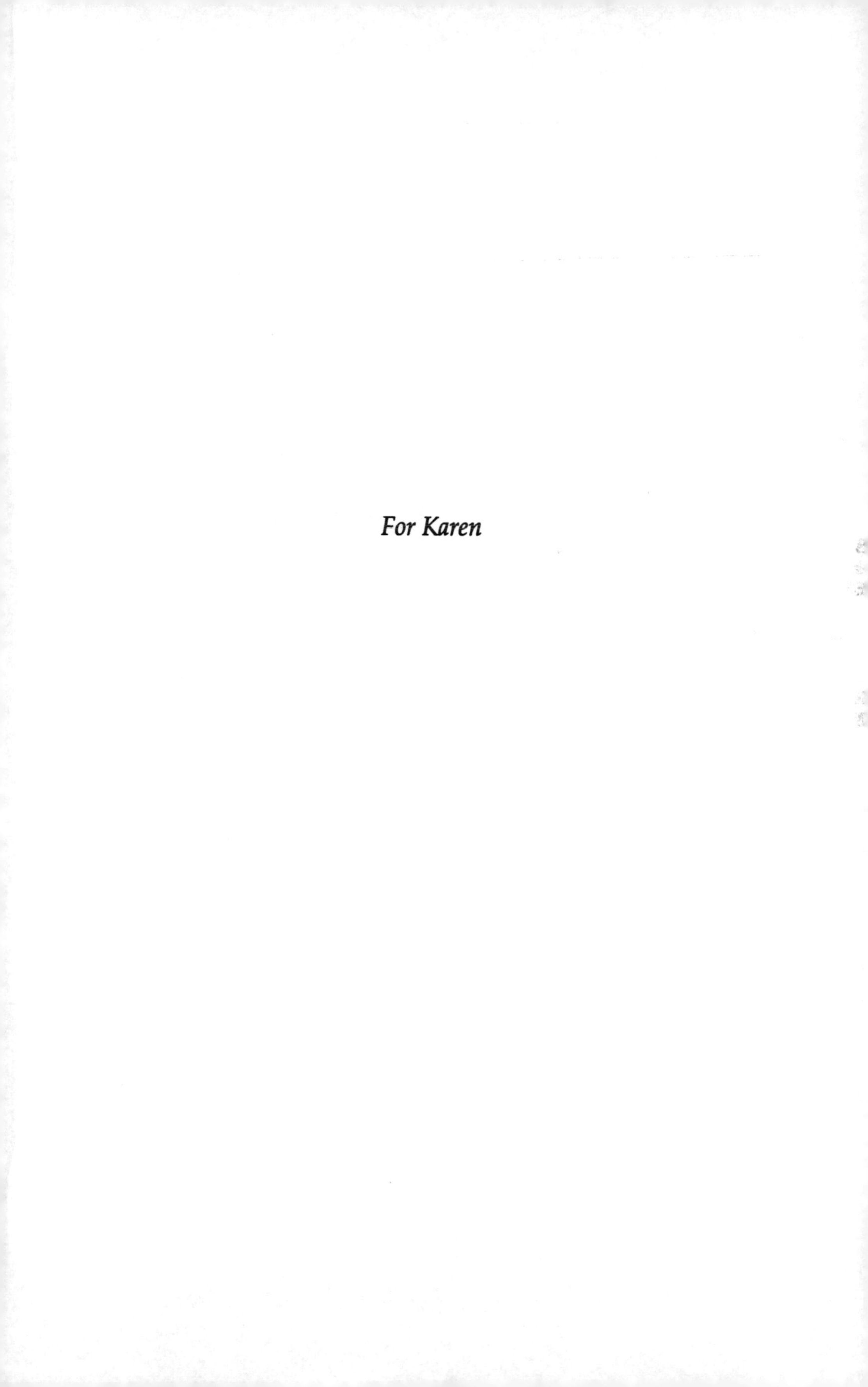

For Karen

Foreword

We are all apprentices of a craft where no one ever becomes a master.
—Ernest Hemingway

In October 1984, the weak writing in a scientific report made national news. The report, which outlined safety procedures during a nuclear attack, advised industrial workers "to don heavy clothes and immerse themselves in a large body of water." The logic behind this advice was sound: Water is a good absorber of heat, neutrons, and gamma rays. Unfortunately, the way the advice was worded was unclear. Was everyone supposed to be completely submerged? Was it safe to come up for air? Besides being unclear, the writing conveyed the wrong impression to the public. The report came across as saying "go jump in a lake"—not the impression you want to give someone spending thousands of dollars to fund your research.

Chances are that Dan Rather will not quote sentences from your documents on national television, no matter how weak the writing is. Still, your writing is important. On a personal level, your writing is the principal way in which people learn about your work. When you communicate well, you receive credit for that work. When you do not communicate well or are too slow to communicate, the credit often goes to someone else. On a larger level, your writing and the writing of other scientists and engineers influences public policy about science and engineering. When the scientific community communicates well, its opinions shape this public policy. When the scientific community does not communicate well, other groups dictate this public policy.

Although scientific writing is important, many scientists and engineers have never sat down and thought out exactly why they write or what they want their writing to accomplish. Instead, these authors rely on a set of vague conceptions that they have developed over the years. Often these conceptions arise from two untrustworthy sources: simplistic rules and weak examples.

The simplistic rules that scientists and engineers remember originate in freshman composition classes taken years ago, late night conversations with colleagues, DOS and DON'TS articles cut out of company newsletters.

Use synonyms for variety.
Never use the first person.
Always write in the active voice.

These rules contain absolutes such as "always" and "never." Worse yet, many of these rules are untrue. When applied to the wide range of writing situations in science and engineering, these rules fail. Face it: Writing scientific documents is difficult, much too difficult to be solved by a list of one-liners.

An even bigger influence on how scientists and engineers write comes from examples that they read. Just as hearing a spoken dialect influences the way you speak, reading a certain writing style influences the way you write. Word choices, sentence rhythms, even the ways that papers are organized are absorbed by readers. Unfortunately, many writing examples in scientific literature are weak. In many documents, the results are not emphasized well, the language is needlessly complex, and the illustrations do not mesh with what is written.

Because of simplistic rules and weak examples in scientific writing, many conceptions that scientists and engineers have about scientific writing are really misconceptions. For instance, many authors think of scientific writing as a mystical aspect of science. Scientific writing is not that at all. For one thing, scientific writing is not a science. It does not contain laws obtained through derivations or experiments. Scientific writing is a craft. It consists of skills that are developed through study and practice. Moreover, scientific writing is not mystical. In fact, scientific writing is straightforward. Unlike other forms of writing, such as fiction, where the goals are difficult if not impossible to define, scientific writing has two specific goals: to inform readers and to persuade readers.

How do you achieve these goals? When the purpose of the writing is to inform, you write in a style that communicates the most amount of information in the least amount of reading time. When the purpose of the writing is to persuade, you write in a style that presents logical arguments in the most convincing manner. You should understand, though, that there are no cookbook recipes for these styles—the writing situations in science and engineering are just too diverse for recipes to apply.

If this book doesn't give recipes, what does it do? First, this book dispels the common misconceptions that prevent scientists and engineers from improving their writing. Second, this book uses examples from actual documents to show the differences between strong scientific writing and weak scientific writing. Third, this book discusses the style of scientific writing by going beyond the surface question of *how things are written* to the deeper question of *why things are written as they are*. In essence, what this book does is make you a critical reader of scientific writing so that you can craft a style for your own writing situations.

In addition to discussing the style of scientific writing, this book discusses the act of sitting down at the computer to write: getting in the mood, writing first drafts, revising, and finishing. I wish that I could tell you that this book will make your scientific writing easy. Unfortunately, that's not the way scientific writing is. Scientific writing is hard work. The best scientific writers struggle with every paragraph, every sentence, every phrase. They must write, then rewrite, then rewrite again. Scientific writing is a craft, a craft you continually hone.

Michael Alley
Madison, Wisconsin
May 1995

Acknowledgments

In preparing this edition, I owe much to my colleagues, especially Leslie Crowley at the University of Illinois, Laura Grossenbacher at the University of Wisconsin, Christene Moore at the University of Texas, and Harry Robertshaw at Virginia Tech. To my students I owe even more, especially those at Virginia Tech, the University of Wisconsin, the University of Texas, Sandia National Laboratories, and Lawrence Livermore National Laboratory. They have challenged my advice, revised it, made it more precise.

Contents

Chapter 1

Introduction: Deciding Where to Begin

But in science the credit goes to the man who convinces the world, not to the man to whom the idea first occurs.

—Sir Francis Darwin

Many people underestimate how difficult scientific writing is. One aspect that makes scientific writing difficult is the inherent complexity of the subject matter. Some subjects such as eddies in a turbulent flow are complex because they are so random. Other subjects such as the double helix structure of DNA are complex because they are so intricate. Still other subjects such as the quantum orbits of electrons are complex because they are so abstract.

Besides the inherent complexity of the subject matter, a second aspect that makes scientific writing difficult is the inherent complexity of the language. Scientific language is full of specific terms, unfamiliar abbreviations, and odd hyphenations that you don't find in other kinds of writing:

> This paper shows how intensity fluctuations in the frequency-doubled output of a Nd:YAG pump laser affect the signal generation from coherent anti-Stokes Raman spectroscopy (CARS).

"Frequency-doubled," "Nd:YAG," "CARS"—these aren't the kinds of expressions that you run across in the morning paper. They are a part of a different language, the language of lasers. In science and engineering, there are many such languages, and these languages are rapidly evolving. Adding still more complexity to scientific writing are mathematical symbols and equations:

> The burning rate (Ω) of a homogeneous solid propellant is given by the following equation [Margolis and Armstrong, 1985]:
>
> $$\Omega = \frac{\rho}{\alpha}(2\lambda\tau)^{1/2} Le^{n/2}\left[\frac{c(1-\sigma)}{(c-\sigma)(1+\gamma_x)}\right]$$

In addressing these complexities of scientific writing, this book does not offer simplistic advice. Nor does it consider only watered-down problems. Rather, this book presents scores of real-world examples that reveal the differences between strong scientific writing and weak scientific writing. In essence, what this book does is make you a critical reader of scientific writing so that you can revise your own work.

Establishing Your Constraints

Whenever I teach writing to scientists and engineers, I begin by asking how they would solve a technical problem, such as the flight of a launched projectile. Here, each scientist and engineer begins the same way, by defining the problem's constraints. What is the projectile's initial velocity? What is the projectile's shape and roughness? What is the projectile's drag? Then I ask how to begin writing a document about the same technical problem. At this point the answers vary. Some people want to write an outline first. Others want to draft an introduction. Still others want to assemble the illustrations in a storybook fashion. This variation in where to begin writing a document is one reason that so many scientists and engineers struggle with their writing.

So, where should you begin writing a scientific document? Actually, you should begin writing a scientific document at the same place that you begin solving a scientific problem: establishing the constraints. More times than I can count, I've run across instances in which a scientist or engineer at the request of a manager has spent days or weeks writing a report, only to have the first draft come back bleeding in red ink because the author assumed one set of constraints for the document and the manager assumed another. Just as you wouldn't pull out equations to solve a technical problem until you understand the problem's constraints, so too should you not start writing sentences and paragraphs until you understand the document's constraints. In scientific writing, there are four principal constraints to consider: the audience for the document, the *format** for the document, the *mechanics* of the document, and the politics surrounding the document.

In addition to the four imposed constraints, there are two variables that you, the writer, bring to the situation. The first is the work that you want to present. Each area of science and engineering has its own set of terms, symbols, and abbreviations that readers expect you to use. The second variable is the purpose that you have for the document. In scientific writing, you write for two specific goals: to inform readers and to persuade readers. In many documents, you do both. For instance, in a proposal to purchase a certain model of streak camera, you would inform readers about how a streak camera works. In the same proposal, you would persuade readers that one model camera should be chosen over other models.

Audience. The audience of your document determines which words you define, what kinds of illustrations you

*All terms in italics are defined in the Glossary.

use, and how much depth you achieve. Success in scientific writing depends on you making a bridge to your audience. You wouldn't write the same document on gene therapy to a Congressional committee that you would to an international conference of researchers. If you did, you would either overwhelm the Congressional committee or not satisfy the international conference. Because each word and image depends on your audience, defining your audience is the place to start writing the document.

In scientific writing, unlike journalism, the types of audiences vary tremendously. While a newspaper reporter writes each story to essentially the same general audience, a scientist or engineer writes to many different audiences. For example, let's suppose that you used technology from weapons research to design an electronic implant for delivering insulin to diabetics [Carlson, 1982]. You might write one document to other engineers who are familiar with the electronics of your design, but not with its application to combatting diabetes. You might then write a second document to medical doctors familiar with diabetes, but not with the electronics of your design. In yet a third document, your audience might be managers who hold the purse strings for your project, but who are unfamiliar with both diabetes and electronics. Here you have three different audiences and three different documents.

The key word in tailoring your writing to the audience is "efficiency." When the purpose of the writing is to inform, efficiency refers to a style that communicates the most amount of information in the least amount of reading time. Put yourself in the shoes of the audience. Many times, when you read a document, you're interested in only the most important details. Too often, though, scientists and engineers present their work as if they were explaining a murder mystery—saving what they think is best for last. The problem is that many readers never get far enough in the document to see the most important details, and many of those who do get far enough are not able to distinguish the most important details from the avalanche of other details. Efficient writing, though, emphasizes important details by placing those details where they stand out.

When the purpose of the writing is to persuade, efficiency refers to a style that presents logical arguments in the most convincing manner. These two styles (the informative and persuasive) differ at times. For instance, while you state your conclusions up front in purely informative documents, you sometimes withhold your conclusions in persuasive documents, especially when the audience includes readers who are antagonistic to those conclusions. An example situation would be a report to the public that recommends a site for a nuclear waste repository. In such a report, you would be better off beginning at a point that no one disputes (perhaps the importance of safety for the surrounding area) and methodically working from that point through your methods until you arrive at your conclusions. In this situation, by keeping the audience in the dark about your conclusions, you create the opportunity to gain credibility with the audience and to present your case before your opposition tunes you out. Will such a strategy convince everyone to accept your conclusions? No. However, if your arguments are cogent, some readers who were on the fence or mildly against the conclusions might swing over, and some who were vehemently opposed might lose some of their vehemence.

No matter what document you are writing, you should assess the audience: (1) who will read the document, (2) what do they know about the subject, (3) why will they read the document, and (4) how will they read the document. These questions dictate how you write the document. In answering the first question, you assess what types of readers you have. Sometimes you have only one type of reader—managers, for example. In more difficult situations, you have a mixture of audiences: managers, scientists and engineers in your field, and scientists and engineers from other fields.

In answering the second question, you assess how much your audience knows about the document's subject matter. From that answer, you can decide what background information to include and which terms to define. Note that when you have a mixture of audiences, you should assess which readers constitute your primary audience and which readers make up your secondary audiences. That determination is important for the organization of the document. In a formal report, for example, the primary audience you address in the *main text*, and the secondary audiences you address in the *back matter*.

In answering the third question, you discern what information the audience is seeking in the document. From that answer, you can decide which results to emphasize. In scientific writing, it is not sufficient that you just logically organize the details. You also should give appropriate emphasis to the details.

Finally, in answering the fourth question, you assess the manner in which the audiences will read the document. Will they read it from beginning to end or will they read only certain sections? Will they just scan the document or will they methodically read through it? Letters, for example, are read quickly, and for that reason, strong writers keep sentences and paragraphs in a letter relatively short. That style allows the audience to sprint through the letter and yet digest the information.

Format. A second constraint of scientific writing is the document's format. Format is the way the type is arranged on the page. Format includes such things as the typeface used, the way the pages are numbered, the way sources are referenced, and the length of the document.

There is no absolute ordained format for scientific writing. Because most journals, laboratories, and corporations establish their own formats, these formats vary considerably (for some general guidelines regarding format, see Chapter 16). What's important is that you follow the format of your situation. If the format for a patent application asks for a line drawing, you supply one. If the format for a company report asks for an executive summary, you supply one. If the format for a proposal asks for a statement of the problem, you supply one.

Many scientists and engineers fret over the different formats in scientific writing. Why, for example, does *Journal A* use one type of referencing system, while *Journal B* uses another? These scientists and engineers seek absolutes in scientific writing; they mistakenly treat scientific writing as a science instead of a craft. Instead of worrying about format, over which you may have little control, you should worry about *style*, something you do control. For example, you should worry about your word choices, the complexity of your illustrations, and the way you structure your papers. Stylistic decisions determine whether readers understand your work. Stylistic decisions determine the success of your writing.

Mechanics. A third constraint of scientific writing is mechanics. Mechanics encompasses the rules of grammar and punctuation. Because science and engineering are based on logic, many scientists and engineers find mechanics frustrating for several reasons. First, there are many rules. For instance, it's not unusual for a mechanics handbook to devote twenty or more pages to rules governing the comma. Second, the rules of mechanics in English have many inconsistencies. One example is where end quotation marks are placed. In the United States, end quotation marks appear outside of periods and commas. In Great Britain, however, end quotation marks often appear inside periods and commas.

A third reason that many scientists and engineers find mechanics difficult is because of the large amount of gray area. For instance, do you refer to the years between 1990 and 1999 as the 1990s or the 1990's? Actually, this aspect of mechanics is governed by format. In *Scientific American*, for instance, it is proper to write 1990s. In *The New York Times*, 1990's is proper. Finally, many scientists and engineers find mechanics frustrating because the English language is constantly changing. A rule that applied in the 1920s (the expression of Roentgen rays as "X rays") does not apply today (at present, "x-rays" is proper).

No simple advice exists to handle the maze of mechanics rules in the English language. As a writer, you have to distinguish between the hard rules, such as the spelling of the word "receive," and the semi-hard rules, such as forming the possessive of a singular noun by simply adding *'s* (an exception would be "Mount St. Helens' eruption"). For more discussion about mechanics, see Appendix A and Appendix B.

Politics. The fourth writing constraint, the politics surrounding the document, is the most difficult to discuss. In an ideal world, this constraint would reduce to the simple statement that you remain honest. Staying honest is straightforward enough. For instance, if you know that your vacuum pumps have coated your experiment with mercury vapor and you suspect that the vapor has altered the results, then you would be dishonest if you did not state your suspicion.

Unfortunately, the world is not ideal, and scientists and engineers are often constrained in their writing not only by the need to remain honest, but also by the need to satisfy lawyers and bureaucrats. Illustrating this point is the poor communication between engineers and management preceding the space shuttle *Challenger* accident. Well before the accident, engineers at Morton Thiokol

International, a contractor for the shuttle, not only knew about the O-ring erosion in the shuttle's field joints, but also had evidence that lower temperatures would exacerbate this problem [*Report*, 1986]. When temperatures plunged below freezing the night before the fateful launch, the engineers tried to have the launch stopped. However, they were rebuffed by their own management at Thiokol and by NASA officials at Marshall Space Center. Why? One reason was the political atmosphere surrounding the shuttle launches at that time. NASA was under such political pressure to launch the space shuttle, and to launch it often, that agency officials and contractors were hesitant to raise issues that would slow the schedule.

Political pressures frustrate scientists and engineers in their writing. I wish that I could say something that would make these constraints go away, but I can't. In some dramatic cases, such as the space shuttle *Challenger*, the political constraints force you to face your ethical responsibilities as a scientist or engineer and to act accordingly. In most cases, the stakes aren't nearly so high. In these smaller cases, the important thing is that you distinguish between the political constraints imposed on that document and the stylistic goals desired for all the rest of your writing. Just because a company lawyer insists that you be wordy in one paragraph does not mean that wordiness is a desired goal for scientific writing.

Selecting Your Stylistic Tools

If you asked ten scientists and engineers to define the term "heat convection," you would receive essentially the same answer ten times because most scientific and technical terms are universally defined. However, if you were to ask the same ten scientists and engineers to define the writing term "style," you might receive ten different answers. Unfortunately, no universal definitions exist for scientific writing—which is one reason that so much confusion exists about the subject.

In this book, style is the part of writing that you, the writer, control. Style is the way in which you cast your thoughts into words and images. Style includes such things as the way you emphasize details, the sentence lengths you choose, the level of detail you use in your line drawings. Style comprises three elements: structure, language, and illustration.

Structure. Structure is the strategy of a scientific document. When most people think about the structure of a document, they think of the word "organization." Granted, the organization of details within the document is a major part of structure, but structure also includes the depth of details, the transitions between details, and the emphasis of details. Structure is the most important element of your style. When your language or illustration falters, your trail breaks and you trip your readers. However, when your structure falters, your trail ends and you lose your readers.

How do you choose the best structure for a document? Given the wide range of topics and audiences in science and engineering, this question is difficult to answer. Despite that difficulty, many scientists and engineers search for template outlines to handle their writing situations. These authors collect template outlines for formal reports, informal reports, feasibility reports, completion reports, and so on. This approach is flawed. For one thing, who wants to keep track of all those outlines? For another thing, when applied to real writing situations, these template outlines just don't work. Granted, you can point out common structural threads for such documents as correspondence, proposals, and instructions, but to lump all proposals into a single packaged outline doesn't make sense. How can one packaged organization work for both a ten-million-dollar grant to contain nuclear fusion with magnetic mirrors as well as a fifty-thousand-dollar grant to study the effects of insects on artichokes. If the outline does work for both projects, then it's so general that it's not worth memorizing, and if the organization does not work, as is probably the case, then you've sacrificed grant money for the convenience of a paint-by-numbers strategy.

Rather than present paint-by-numbers strategies, this book shows the differences between documents with strong structure and documents with weak structure. In doing so, this book examines structure from four perspectives: the organization of details, the transition between details, the depth of details, and the emphasis of details. For each perspective, strong and weak examples are shown.

Language. Language is the way that words are used. Language is not only the choice of words, but also the arrangement of those words in phrases and sentences. In scientific writing, language includes the use of numbers, equations, and abbreviations; it includes the use of examples and analogies.

In scientific writing, your language must first be precise. You must say what you mean. Your language must also be clear. While precision means saying what you mean, clarity means avoiding things that you don't mean. Besides being precise and clear, you should be forthright in your language. When you write a scientific document, you assume the role of a teacher, and as a teacher you want to convey a sincere and straightforward attitude. Another goal of language is to anchor your language in the familiar. In other words, you should use language familiar to your readers. Before readers can learn anything new, they have to see it in relation to something they already know.

Moreover, because scientists and engineers produce so much writing, they have an obligation to keep their language concise. Every word should count. Although you should make your language concise, you should also make it fluid. Fluid writing is smooth writing, writing with transition, writing that moves from sentence to sentence, paragraph to paragraph without tripping or tiring the reader.

These six goals for language do not all carry the same weight. Precision and clarity are the most important. The relative importance of the other goals depends on your audience and work. For instance, in an abstract paper about quantum mechanics, being familiar might become more important than being concise. In this paper, you could add analogies to help readers understand your ideas.

Don't assume that these six goals for language are in constant conflict. Most of the time, these goals reinforce one another, as shown in Figure 1-1. When you are clear and forthright, conciseness follows. When you are precise and familiar, clarity follows. Also, don't think that by pursuing these six goals you will lose your individu-

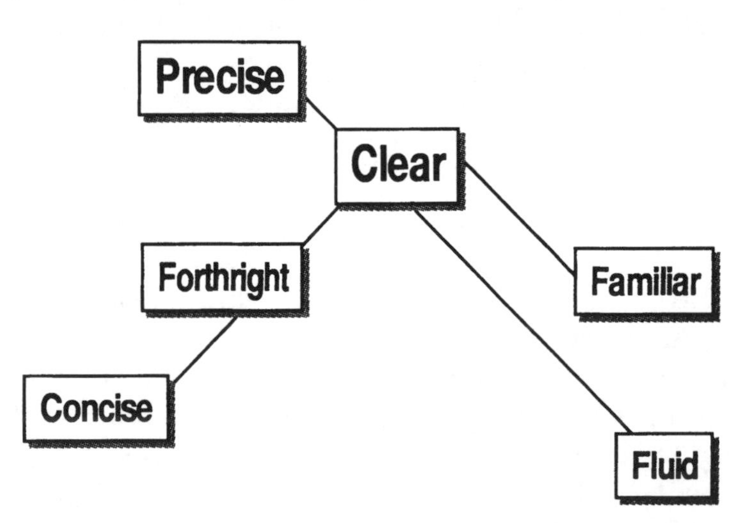

Figure 1-1. Six goals of language in scientific writing. Precision and clarity are the most important.

ality as a writer. Within these six goals there is much variation. Pursuing these language goals will not cause your writing to sound like everyone else's; rather it will make your writing succeed.

Illustration. Illustration is not just the presence of figures and tables in a document. Illustration is the meshing of figures and tables with the language. Just pasting pictures into your paper is not strong scientific writing. In fact, a slapped-in figure or table may confuse your readers more than inform them.

Other types of writing, such as literary novels, do not generally use illustrations. Why do we use illustrations in scientific writing? Unlike the purpose of fiction, which is difficult to define, the purpose of scientific writing is straightforward: either to inform or to persuade the audience as efficiently as possible. Illustrations can help make your writing efficient by clarifying images that are too complex to be conveyed by language. For instance, just telling your audience that the eruption of Mount St. Helens in 1980 had an ash plume that reached 19,000 meters and dispersed 0.67 cubic kilometers of material into the atmosphere is not nearly as effective as anchoring that data with an illustration such as Figure 1-2. Note that just having the figure without the accompanying data would not have been effective because the readers would not have known the heights and amounts involved.

Besides making the reading more efficient, illustrations can also make the writing more efficient. Imagine presenting the information in Table 1-1 in sentence form. Without the table, you would end up writing strings of tedious sentences that would simply list measurements and ratios. Also, without the table, your readers would waste much time sifting through these sentences to find specific details such as the gravity on Jupiter.

Figure 1-2. Eruption of Mount St. Helens in 1980 (courtesy of the United States Geological Survey).

Table 1-1
Characteristics of Planets in Our Solar System [*CRC Handbook*, 1995]

Planet	Radius (kilometers)	Gravity (m/s^2)	Year (earth days)
Mercury	2,440	3.70	88
Venus	6,052	8.87	225
Earth	6,378	9.80	365
Mars	3,397	3.71	687
Jupiter	71,492	23.12	4,333
Saturn	60,268	8.96	10,759
Uranus	25,559	7.77	30,685
Neptune	24,764	11.00	60,181
Pluto	1,151	0.72	90,470

Although illustrations can make your writing more efficient, you shouldn't fill your journal articles and formal reports with graphics in the same way that a comic book is filled with pictures. Too many illustrations will reduce the emphasis given to any one illustration and undercut the efficiency of informing. As with language, the way you illustrate depends on your constraints.

References

Carlson, Gary, "Implantable Insulin Delivery System," *Sandia Technology*, vol. 6, no. 2 (June 1992), pp. 12–21.

CRC Handbook of Chemistry and Physics, 75th ed. (New York: Chemical Rubber Publishing Co., 1995), chap. 14, p. 3.

Darwin, Sir Francis, "Sir Galton Lecture Before the Eugenics Society," *Eugenics Review*, vol. 6, no. 1 (1914).

Hemingway, Ernest, interview, *New York Journal-American* (11 July 1961).

Report of the Presidential Commission on the Space Shuttle Challenger Accident, vol. 1 (Washington, DC: White House Press, 1986), pp. 104–110, 249.

Chapter 2

Structure: Organizing Your Documents

If a man can group his ideas, then he is a writer.
—Robert Louis Stevenson

When discussing the organization of documents, Aristotle said, "A whole is that which has a beginning, middle, and ending." This approach is a good way to examine the organization of general scientific documents, such as reports and articles.

The beginning of a report or article serves a specific purpose—it prepares readers for the middle, which is the discussion of the work. In preparing readers for the middle, the beginning fulfills certain expectations of the readers. These expectations include defining the work, showing why the work was done, giving background for understanding the work, and revealing how the work will be presented. The middle, often called the discussion, simply presents the work. The middle states what happened in the work and states how it happened. The middle presents the results, shows where they came from, and explains what they mean. The ending of a scientific document then further analyzes the work presented in the middle and gives a future perspective. While the middle presents each result separately, the ending looks at the results from an overall perspective.

Beginnings of Documents

The beginning to a scientific document has one task: to prepare readers for understanding the document's middle. The beginning to a scientific document is important because it determines whether the audience will continue reading. In a sense, the beginning is a make-or-break situation. The beginning of a scientific document includes the title, summary, and introduction.

Creating Titles. The title is the single most important phrase of a scientific document. The title tells readers what the document is. If your title is inexact or unclear, many people for whom you wrote the document will never read it. Consider a weak example:

Reducing the Hazards of Operations

What is this document about? Only a psychic could know. This document could be about anything from using catalytic igniters in a nuclear power plant to using new plastic gloves during operations on AIDS patients. Would you search the library stacks for this document? Probably not—your time is too valuable to spend on such a search.

Ideally, a strong title to a report or article orients readers in two ways: first, it identifies the field of study for the document; and second, it separates the document from all other documents in that field. A good test for a title is the way it reads in a list of titles recovered by a computer search. A strong title will meet the two criteria; a weak title will not.

A strong title identifies the field of study for the work. Consider an example that does not succeed:

Effects of Humidity on the Growth of Avalanches

Although this title is more specific than the first example, it still does not meet the first criterion. On the basis of this title, you might assume that this document is a geological study of rock or snow avalanches in a mountainous terrain. Actually, this document is about electron avalanches in electrical gas discharges. Therefore, the title should be revised to reveal the field of study:

Effects of Humidity on the Growth
of Electron Avalanches in Electrical Gas Discharges

Just because a title names the field of study does not mean the title is strong. Naming the field of study gets your audience to the right ballpark, but your audience still doesn't know who the teams are. In other words, you have to address the second criterion, which is to separate your work from everyone else's. Consider the following example:

Studies on the Electrodeposition of Lead on Copper

Although this title orients the audience to the area in which the work was done—plating of lead onto copper—this title is still unsuccessful. Somehow, the writer has to distinguish this work from other work in the area. For the work that this paper discussed, a better title would have been

Effects of Rhodamine-B on the Electrodeposition
of Lead on Copper

Now this title orients readers to the area of work and gives a specific detail ("effects of rhodamine-B") to distinguish this work from other work in the area.

In a title, an audience can absorb only three or four details. More than that—things begin to blur. For that reason, giving too many details is as weak as giving too few:

Effects of Rhodamine-B and Saccharin
on the Electric Double Layer During
Nickel Electrodeposition on Platinum
Studied by AC-Cyclic Voltammetry

There's just too much information given here. In a strong title, you must balance each detail's contribution against the space it acquires. If the principal aspect is the use of the new technique (AC-cyclic voltammetry), then a stronger title would be

Use of AC-Cyclic Voltammetry to Study Organic Agents
in the Electrodeposition of Nickel on Platinum

Without being too long, this title emphasizes the unique element of the research: AC-cyclic voltammetry. Ideally, you want your title to identify your work so that it stands apart from any other work on your mountain. Often, though, you cannot achieve this goal in a phrase that is both clear and precise. Nonetheless, you can usually find a title that indicates the most distinctive aspect of your work. What about details of secondary importance? Those you can present in the summary or introduction.

When writing titles, many scientists and engineers fall in love with big words and forget about the importance of the small words that are needed to couple those big words. Unfortunately, strings of big words are difficult to read.

10 MWe Solar Thermal Electric Central Receiver Barstow Power
Pilot Plant Transfer Fluid Conversion Study

What is this report about? Perhaps we can guess that some kind of solar energy plant is involved. But does this title orient? No, it overwhelms. Many details are included, but we have no sense of the relationship of those details. This particular document proposed a new heat transfer fluid for Solar One, the world's largest solar power plant. Given that, a stronger title would have been

Proposal to Use a New Heat Transfer Fluid
in the Solar One Power Plant

Notice how this revised title contains short words—"to," "a," "in," and "the"—interspersed among the bigger words. These smaller words serve as rest stops for the audience. Notice also that this document was a special situation, a proposal, as opposed to the typical situation of a report or article. By identifying special situations such as proposals and instructions in the titles, you orient the audience to the specific perspectives of those documents.

When writing a title, you should consider your readers. What do they know about the subject? What do they not know? In a title, avoid phrases that your audience will not recognize. If readers do not understand the title, they will often not read any further in the document.

Use of an IR FPA in Determining
the Temperature Gradient of a Face

Although most readers will realize that the engineer determined temperature gradients in this work, most readers probably will not recognize the acronym "IR FPA." Perhaps readers might realize that IR stands for infrared, but what about FPA? Another problem with this title is the ambiguous use of the word "face." What kind of face? A crystal face? A mountain face? In this engineer's case, the face was actually a human face, a detail that in the work was relatively unimportant. What was important here was that the engineer had developed a new way to measure temperature gradients. For that reason, a better title would have been

Determining Temperature Gradients
With a New Infrared Optical Device

In this revision, you give enough information to orient, but not so much information that you confuse.

Writing Summaries. Winston Churchill said, "Please be good enough to put your conclusions and recommendations on one sheet of paper at the very beginning of your report, so that I can even consider reading it." When the purpose of the writing is to inform, that is what summaries should do: give away the show right from the beginning and let the audience decide whether they want to read the document. Scientific writing is not mystery writing in which the results are hidden until the end. In most scientific documents, the strategy is to state up front what happened and then use the rest of the document to explain how it happened.

Many scientists and engineers find the principle of summarizing their work at the beginning difficult to swallow. They don't believe that audiences will read their papers and reports all the way through if the results are stated up front. Actually, these authors are right—many readers, after seeing a summary, will not read the entire document. However, the readers who are truly interested in the work will continue reading. Remember: The goal of scientific writing is not to entice all audiences to read to the end of your document, but either to inform or persuade the audience as efficiently as possible.

Besides emphasizing the most important details, summaries also make it easier for audiences to read through complex documents. Not being told what is going to happen in a complex document is akin to being blindfolded and forced to hike a difficult trail. Because you aren't sure in which direction you're headed or how far you'll be going, you're ready to quit as soon as the trail gets rough. The same is true for a document that doesn't state its results up front. For instance, in a paper filled with Monte Carlo simulation techniques, you may tire if you don't know what those simulations accomplish. If, however, you know that those simulations shed new light on nuclear fusion reactions, then you might stay with the paper.

Although there are many names for summaries in scientific writing, there are two main types: descriptive summaries and informative summaries. The descriptive summary tells what kind of information will occur in the document; it is a table of contents in paragraph form. The informative summary tells what results occurred in the work; it is a synopsis of the work. Not all summaries are entirely descriptive or informative. Many summaries are combinations of the two. Nevertheless, it's important to understand these two types and the advantages and disadvantages of each.

A descriptive summary (sometimes called a descriptive abstract) tells readers what kind of information the document will contain. The descriptive summary is like the by-line to a baseball game:

New York (Seaver) versus Baltimore (Cuellar)

From the byline, you know what's going to happen—which teams will play and who will be the pitchers. Descriptive summaries give the same kind of information about the document, namely, what the document will cover:

A New Chemical Process for Eliminating Nitrogen Oxides From Diesel Engine Exhausts

> This paper introduces a new chemical process for eliminating nitrogen oxides from the exhausts of diesel engines. The process uses isocyanic acid, a nontoxic chemical used to clean swimming pools. In this paper, we show how well the process reduced emissions of nitrogen oxides from a laboratory diesel engine. To explain how the process works, we present a scheme of chemical reactions.

Note that the first sentence of this descriptive summary orients the audience to the identity of the work. Don't think that the repetition between the title and summary is redundant. Being redundant is a needless repetition of details. The repetition here is purposeful—you want to clarify any doubts that the audience has about the meaning of the title. Notice also that the second sentence of this descriptive summary provides a secondary detail that could not fit into the title. When you cannot find a concise title to separate your work from everyone else's work, you can use the summary to do so. The final two sentences of this descriptive summary list chronologically what will occur in the document: a discussion of the experiment followed by discussion of the theory.

Notice also that a descriptive summary can be written ahead of time. Because the descriptive summary tells what the document will cover, instead of which results were found, you can often write the descriptive summary days, weeks, even months before the document. In fact, many people find themselves writing descriptive summaries to conference proceedings, even though the work is not yet finished. Notice also how concise a descriptive summary is, often only two or three sentences. For that reason, this kind of summary can be read quickly:

> This paper describes a new inertial navigation system for mapping oil and gas wells. In this paper the mapping accuracy and speed of this new system are compared against the mapping accuracies and speeds of conventional systems.

An informative summary, often called an executive summary when written for management, is the second kind of summary. Unlike descriptive summaries, informative summaries do present the actual results of the work. Informative summaries are analogous to box scores from baseball. In box scores, such as the ones for the famous 1969 World Series shown below, you gather the most important results of the games: how many runs (R) each team scored, how many hits (H) each team had, and how many errors (E) each team made. You also gather many secondary results such as who the winning and losing pitchers were and who hit home runs.

	First Game			R	H	E
New York Mets	0 0 0	0 0 0	1 0 0	1	6	1
Baltimore Orioles	1 0 0	3 0 0	0 0 x	4	6	0

Winning pitcher—Cuellar (1–0). Losing pitcher—Seaver (0–1).
Home runs—Baltimore: Buford (1).

Second Game					R	H	E
New York Mets	0 0 0	1 0 0	1 0 1		2	6	0
Baltimore Orioles	0 0 0	0 0 0	1 0 0		1	2	0

Winning pitcher—Koosman (1–0). Losing pitcher—McNally (0–1).
Home runs—New York: Clendenon (1).

Third Game					R	H	E
Baltimore Orioles	0 0 0	0 0 0	0 0 0		0	4	1
New York Mets	1 2 0	0 0 1	0 1 x		5	6	0

Winning pitcher—Gentry (1–0). Losing pitcher—Palmer (0–1).
Home runs—New York: Agee (1) and Kranepool (1).

Fourth Game					R	H	E
Baltimore Orioles	0 0 0	0 0 0	0 0 1	0	1	6	1
New York Mets	0 1 0	0 0 0	0 0 0	1	2	10	1

Winning pitcher—Seaver (1–1). Losing pitcher—Hall (0–1).
Home runs—New York: Clendenon (2).

Fifth Game					R	H	E
Baltimore Orioles	0 0 3	0 0 0	0 0 0		3	5	2
New York Mets	0 0 0	0 0 2	1 2 x		5	7	0

Winning pitcher—Koosman (2–0). Losing pitcher—Watt (0–1).
Home runs—Baltimore: McNally (1), F. Robinson (1); New York: Clendenon (3), Weiss (1).

From these box scores, you can infer that the Mets won the series (four games to one), that Clendenon hit three home runs, and that Koosman, Gentry, and Seaver recorded wins for the Mets. Informative summaries give you the same kind of information—namely, what happened in the project.

Like the descriptive summary, the informative summary begins by identifying the project, but then informative summaries go much further. Informative summaries also state the main results of the project. In essence, informative summaries give readers the major conclusions and recommendations at the beginning of the document. Consider an example [Kelsey, 1983]:

> This paper describes a new inertial navigation system that will increase the mapping accuracy of oil wells by a factor of ten. The new system uses three-axis navigation that protects the sensors from high spin rates. The system also processes its information by Kalman filtering (a statistical sampling technique) in an on-site computer. Test results show that the three-dimensional location accuracy is ±0.1 meters

> of well depth, an accuracy ten times greater than conventional systems.
>
> Besides mapping accuracy, the inertial navigation system has three other advantages over conventional systems. First, its three-axis navigator requires no cable measurements. Second, probe alignment in the borehole no longer causes an error in displacement. Third, the navigation process is five times faster because the gyroscopes and accelerometers are protected.

This informative summary is tight—there is no needless information. Informative summaries are a sum of the significant points, and only the significant points, of the project. Informative summaries are also independent of the paper itself. For instance, unusual terms, such as Kalman filtering, are defined. After reading the informative summary, the audience would read the main text of the document to find out how the work was done, not what happened.

Everything written in the informative summary—every sentence and illustration—is either a repetition or condensation of something in the main text of the document. Because informative summaries are drawn from the main text of the document, they are the last section written. Typically, informative summaries are about 5 to 10 percent of a document's length. In a formal report, that 5 to 10 percent may include illustrations.

Which type of summary should you use? Descriptive or informative? Sometimes the purpose and audience dictate which type to use. As stated in Chapter 1, if the purpose of the document is to persuade and if you have an audience antagonistic to your results, you would not state your results up front. For instance, assume that after a long study you have decided to allow a company to mine zinc in an environmentally sensitive area. In the report that announces this decision to the public, you would withhold your decision until the latter part of the report so that you can first present your arguments for that decision. In such a report, you would use a descriptive summary rather than an informative summary.

Format often dictates which type(s) of summary you use. Journals, for example, have such tight length restrictions that you seldom have room to write an informative summary. In some journals, you have room for only a descriptive summary. Formal reports, on the other hand, do not have such tight restrictions. For that reason, you often have room in formal reports to include both types of summaries.

In still other situations, you have room for something in the middle—a summary that blends informative and descriptive elements. Consider an example [Perry and Siebers, 1986]:

A New Chemical Process for Eliminating Nitrogen Oxides From Engine and Furnace Exhausts

> This paper introduces a new chemical process for eliminating nitrogen oxides from engine and furnace exhausts. Nitrogen oxides are a major ingredient of smog and contribute heavily to acid rain. In this process, isocyanic acid—a nontoxic chemical used to clean swimming pools—converts the nitrogen oxides into steam, nitrogen, and other harmless gases. While other processes to reduce nitrogen oxides are expensive and, at best, only 70 percent effective, our new process is inexpensive and almost 100 percent effective.
>
> In laboratory tests, our process eliminated 99 percent of nitrogen oxides from the exhaust of a small diesel engine. If incorporated into diesel engines and industrial furnaces, this new process could greatly reduce the 21 million tons of nitrogen oxides released each year into the atmosphere of the United States. Besides presenting experimental results, this paper also presents a scheme of chemical reactions to explain how the process works.

Most of the sentences in this summary are informative. These sentences present the most important results: the description of the new process and its effectiveness at reducing nitrogen oxides from the exhaust of a test engine. The last sentence of the summary, though, is descriptive.

Instead of actually presenting the chemical reactions that explain the process, the summary just states that the scheme will be given. Such a descriptive treatment was necessary because the format didn't allow room for all six chemical equations.

Writing Introductions. When audiences read an introduction to a scientific document, they have expectations that have arisen from reading the introductions of other scientific documents. In general, by the end of the introduction, audiences expect answers to the following questions:

What exactly is the work?
Why is the work important?
What is needed to understand the work?
How will the work be presented?

Don't assume that your introduction must explicitly address all four questions. Depending on the work and the audience, your introduction may address only two or three of the questions. Also, don't assume that for every document the most efficient order for answering the questions is the one listed above. Again, the way you write your introduction depends on your work and your audience. In one document, you may begin your introduction by explaining what the work is. On another document, though, you may feel that your audience needs some background before learning the identity of the work. Although introductions vary in the type and order of information, introductions should be designed so that readers do not reach the middle of your document with any of these four questions still burning.

Your introduction is your first chance to define the full boundaries of your work. In the introduction, you're not cramped by space as you are in your title and summary. Therefore, you should take advantage of the opportunity:

> This paper presents a model to describe the electrical breakdown of a gas. We call this model the two-group model because of the similarity between the problem of gas breakdown and the problem of neutron transport in nuclear reactor physics. The two-group model is based on electron kinetics and applies to a broad range of conditions (breakdown in pure gases, for example). The model also provides a continuous picture of the initial phase of breakdown above the Townsend regime, both in structure of the breakdown and in the physics of the processes. [Kunhardt and Byszewski, 1980]

This introduction gives details about the work that couldn't fit into the title or summary, details such as where the theory got its name and the theory's relation to other theories.

When you identify your work in your introduction, you should specify the scope and limitations of the work. The scope includes those aspects that the project includes. The limitations are the assumptions that restrict the scope's boundaries. Scope and limitations usually go hand in hand. Often, when you identify a project's scope, you implicitly state what the limitations are. Sometimes, though, you must clarify your limitations:

> In this paper, we have compared the life expectancies of three different groups of people: heavy alcohol drinkers, moderate alcohol drinkers, and people who do not drink alcohol. We have not, however, studied the social, medical, or economic makeup of these groups—three elements that could affect life expectancies much more than alcohol intake.

The first sentence of this example specifies the scope, and the second sentence specifies the limitations. In this example, you have to specify your limitations because your limitations raise important questions that your readers might not have inferred from the scope.

Besides being an opportunity to define your work, the introduction is an opportunity to show why your work is important. Unfortunately, many scientists and engineers launch into the project's nuts and bolts without showing the importance of the work. The result is that many readers don't finish the document because they have no reason to work through the details. Reading scientific documents is taxing work, and readers need incentives to keep going. Showing the importance of the work provides an incentive.

Another reason to show the importance of the work is money. Most scientific projects depend on outside funding, and before someone will give away money, they have to be convinced that the work is important. More often than not, that particular someone will be someone outside science and engineering. Justifying your work to someone outside science can be difficult. You cannot get away with just saying the project is important, as this physicist tried to do:

> This paper presents the effects of laser field statistics on coherent anti-Stokes Raman spectroscopy intensities. The importance of coherent anti-Stokes Raman spectroscopy in studying combustion flames is widely known.

This introduction convinces readers of nothing. Instead of just telling readers that the project is important, you should show readers that the project is important, as this chemist did [Thorne and others, 1985]:

> This paper presents a design for a platinum catalytic igniter in lean hydrogen-air mixtures. This igniter has application in light-water nuclear reactors. For example, one danger at such a reactor is a loss-of-coolant accident, in which large quantities of hydrogen gas can be produced when hot water and steam react with zirconium fuel-rod cladding and steel. In a serious accident, the evolution of hydrogen may be so rapid that it produces an explosive hydrogen-air mixture in the reactor containment building. This mixture could breach the containment walls, allowing radiation to escape. To eliminate this danger, one proposed method is to ignite intentionally the hydrogen-air mixture at concentrations below those for which any serious damage might result.

Although most work has a practical application, don't assume that you have to show a practical application for all work. Many strong projects exist for the sole purpose of satisfying curiosity. In such cases, you cannot assume that your readers already share your curiosity. You must instill that curiosity. You should raise the same questions that made you curious when you began the work:

> In size, density, and composition, Ganymede and Callisto (Jupiter's two largest moons) are near twins: rock-loaded snowballs. These moons are about 5000 kilometers in diameter and contain 75 percent water by volume. The one observable difference between them is their albedo: Callisto is dark all over, while Ganymede has dark patches separated by broad light streaks. This paper discusses how these two similar moons evolved so differently. [LLNL, 1985]

How much space should you devote to justifying your work? That answer depends on your audience. If your readers are experts in your field, you may not have to justify your work explicitly—your readers might implicitly understand the importance. However, not justifying your work limits your audience. Your audience, in essence, becomes only those experts.

The third question that readers expect an introduction to answer (what is needed to understand the document?) is really a question of what background information the introduction provides. That answer depends on your readers and how much they know about your work. For instance, if you were writing about the effects of a long-duration space mission on the human immune system and if your audience was a general scientific and engineering audience, then much of the background would be on the human immune system itself. However, if your audience consisted of immunologists, then much of the background would be on something else, perhaps a review of the immunology findings from previous space missions.

In general, the less your audience knows about your subject, the more difficult it is to write the background section. Unless you plan to spend the rest of your career on one document, you can't begin at the lowest stratum of science with Euclid or Archimedes and cover everything in between. You have to be selective. For instance, if it were 1913 and you were Niels Bohr writing the theory of the hydrogen atom, you might assume your readers were familiar with Balmer's equation for wavelength and Coulomb's law of force, but not with Rutherford's nuclear model for the atom, which was proposed in 1911. You might then start your paper at an "elevation of knowledge" somewhere just below Rutherford's work.

No matter how much your readers know about your work, you should be selective with background material, particularly in journal articles. Because most formats for journals have tight space constraints, you should provide background on those things that your audience really needs. Many scientists and engineers mistakenly assume that they have to provide a historical discussion with each document. If a historical discussion serves your readers in the document, then provide it. However, in many documents, other kinds of information such as definitions of key terms are more important.

Also, don't assume that all background information must go into the introduction. Sometimes, if you have a lot of background information, your document will read more efficiently if in the introduction, you restrict yourself to background that applies to the entire document. In other words, if the background is pertinent to only one section of the middle, then place that background within that particular section. If you have a lot of overall background information, you might place that background information in a separate section following the introduction so that the background information does not overwhelm the other aspects of the introduction.

The last expectation that an audience has for an introduction is the mapping. In general, the longer a document is, the more important the mapping of the work becomes. This principle is not only true in scientific writing, but in all kinds of communication. Anyone who has ever attended a Southern Baptist revival understands this point. In a Southern Baptist revival, the preacher has no time limit. One saving grace, though, is that most Southern Baptist revival preachers use three-point sermons. In a three-point sermon, the preacher states in the beginning the three points to be covered—say Sin A, Sin B, and Sin C—and then the preacher covers those three sins, one at a time, and in the order stated. Once the preacher has covered all three sins, the sermon is over and you sing the invitational hymn. This mapping of the sermon's structure allows the congregation to know at any given moment in the sermon about how much longer the preacher will be speaking. For a congregation in the South during a muggy summer evening, that information is important. If the preacher's only on Sin A, you know you've got a while to go. You sit still and breathe slowly. However, if the preacher is on Sin C, you relax a little, wipe the sweat from your forehead, and slide your thumb to the hymn of invitation.

Although the subject matter for your documents will be different than the subject matter of a revival sermon, the principle of mapping remains the same. Consider the mapping in this journal article about a "nuclear winter" [Garberson, 1985]:

> This report discusses the effects of smoke on the earth's climate following a large-scale nuclear war. In the first section of the report, we present a war scenario in which 10,000 megatons of high-yield weapons detonate. The second section of the report then introduces assumptions for the amount of smoke produced from resulting fires, the chemical characteristics of the smoke, and the altitudes at which the smoke initially enters the atmosphere. In the third section, we present computer models that show how the smoke distributes itself in the weeks and months following the war. Finally, in the fourth section, we discuss how the earth's climate changes as a result of that smoke distribution.

Once you have given a map of your strategy, you are obligated to stick with it. Nothing makes a congregation more restless than a preacher who promises to talk about three sins and then covers four.

You might ask what is the point of mapping the document in the introduction, when the summary has already done that. Two reasons exist. First, by mapping the document at the end of the introduction, you make a nice transition from the beginning to the middle of the document. Second, in some documents, the reader desires a justification for why you organized the document as you did. For instance, in an evaluation article, why do you discuss Option A before Option B? A summary does not have space to provide this kind of information; an introduction does.

Middles of Documents

The middle, or discussion, of a scientific document simply presents the work. In the middle, you state what happened as well as how it happened. You state the results, show where they come from, and explain what they mean. What organization problems must you surmount in the middle? In writing the middle, you select a strategy and then convey that strategy to the audience in your choice of headings and subheadings. There are many logical strategies in scientific writing: chronological strategies, spatial strategies, flow strategies, as well as the traditional strategies, such as cause-effect, that you studied in high school. The names of these strategies aren't so important. What's important is that you choose a logical strategy that is appropriate for your audience. Also important is that you reveal that strategy through your headings and subheadings.

Choosing an Appropriate Strategy. To describe your work, you can draw from a number of strategies. Which strategy is the most appropriate? This answer depends, as you might imagine, on the subject and audience.

One of the most common strategies is the chronological strategy, which follows the variable of time. Chronological strategies are appropriate in discussions of time-line processes and cyclic processes. In a time-line process such as the evolution of Hawaiian volcanoes, you would follow the development of the volcano through its eight life stages. In a cyclic process, such as the orbit of a comet, you would designate a beginning to the orbit and follow the process until it completes the cycle.

Note that in both situations, you assign markers that divide the process into stages or steps. When dividing a process into steps, you should try to group the steps into clusters of twos, threes, or fours so that your audience can remember them. Although readers occasionally remember longer lists, such as the months of the year, lists longer than four tax the memory. How then would you handle a situation, such as the Hawaiian volcano, in which you have eight steps [Bullard, 1976]? One way is to break the list of eight into two lists of four: the building stages and the declining stages. In the building stages, the volcano develops from the sea floor to a volcano above sea level (Kilauea is a good example of a volcano in the building stages). In the declining stages, the volcano deteriorates due to erosional effects (Oahu exemplifies a volcano in the declining stages).

- Building Stages
    - Explosive Submarine Stage
    - Lava-Producing Stage
    - Collapse Stage
    - Cinder-Cone Stage
- Declining Stages
    - Marine and Steam-Erosion Stage
    - Submergence and Fringing-Reef Stage
    - Secondary Eruptions and Barrier-Reef Stage
    - Atoll and Resubmergence Stage

Another common strategy is the spatial strategy. Here, the strategy follows the physical shape of a form or object: the curvature of a fossil, the dispersion of a volcano's ash plume, or the shape of a comet. As with chronological strategies, you would want to divide the form into two, three, or four distinct parts. For instance, in describing a comet, you might divide the comet into its head, coma, and tail.

A third common strategy in science and engineering is to follow the flow of some variable, such as energy or mass, through a system. Consider, for example, the nuclear fusion experiment in Figure 2-1. For this system, you could choose a spatial strategy in which you begin with the circumference of the experiment and then work your way radially inward. Given the complexity of this experiment, though, this strategy proves cumbersome. A better strategy would be to follow the flow of energy through the system. This strategy reduces the experiment from three dimensions to one dimension. In this strategy,

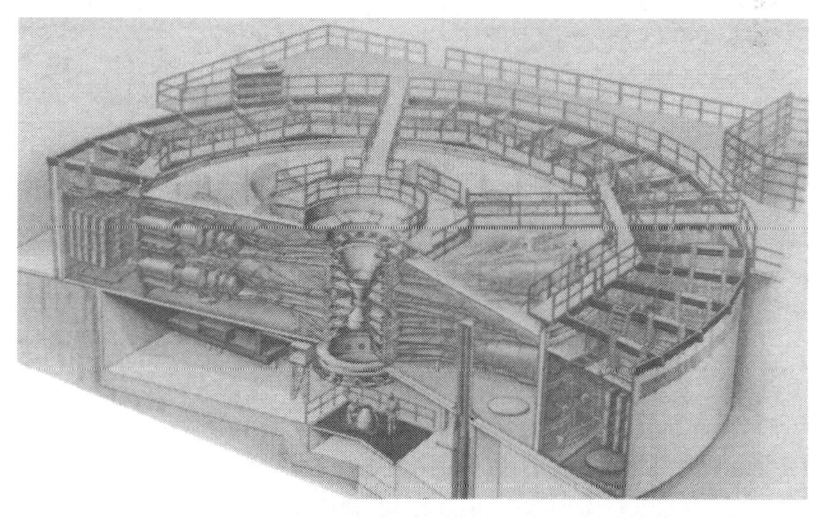

Figure 2-1. Cutaway of nuclear fusion experiment at Sandia National Laboratories. Here, an accelerator focuses lithium ions onto deuterium-tritium pellets in an attempt to produce nuclear fusion [VanDevender, 1985].

you follow the energy as it changes from electrical energy to particle beam energy and then to fusion energy. As with the first strategy, you end up moving radially inward, but unlike the first strategy, you have your audience thinking in one dimension rather than three.

As stated earlier, the organizational problems in scientific writing usually don't arise because the chosen strategies are illogical. Rather, the problems arise because the logical strategies chosen are inappropriate for the audience. In describing the nuclear fusion experiment, the flow of energy from electrical energy to fusion energy works well for a technical audience. However, for a non-technical audience, such as the United States Congress, which is deciding whether to fund this project, you might reconsider this strategy. Because Congress thinks of the project as a nuclear fusion project, you might begin with the fusion energy rather than with the electrical energy:

> In our scheme of producing nuclear fusion, we compress tiny deuterium-tritium pellets, which are about the size of BBs. To compress these pellets, we require the energy of a focused beam—in our scheme, a beam of lithium ions. Producing this particle beam requires a huge pulse of electrical energy, which is supplied here by a Marx bank generator.

For this non-technical audience, we've moved backwards from the recognized goal of producing nuclear fusion to the unfamiliar steps of generating a particle beam and charging a Marx bank generator. We've chosen not only a logical strategy, but an appropriate strategy.

The traditional strategies that you learned in high school are also common in scientific writing. For instance, you use a division-classification organization to group items into parallel parts. Take the example of the global climatic effects following a nuclear blast. These effects include radiation fallout, nitrogen oxides, and smoke. Now you could choose a chronological strategy and discuss each of these three effects during the first week after the blast, then the second week, and then the third. This strategy proves cumbersome though because the time scales for the effects are so different—the global climatic effects of radiation fallout take place over a matter of hours, while the effects of smoke continue for weeks. A better strategy involves treating each effect separately:

Effects of Radiation
Effects of Nitrogen Oxides
Effects of Smoke

Granted, within each of the three sections, you would probably use a chronological strategy, but your overall strategy would be a classification into parallel parts. Cause-effect organizations and comparison-contrast organizations also occur in scientific writing. Cause-effect organizations serve documents in which you investigate why things occur (for instance, why the *Titanic* sank so quickly). Likewise, comparison-contrast organizations serve documents in which you evaluate a number of options (for instance, an evaluation of lifeboat designs for a cruise ship).

Choosing an appropriate strategy is not a paint-by-numbers decision. You can't pull out a chronological strategy for instructions or a spatial strategy for equipment specifications and expect the strategy to work every time. For each document, you should tailor a strategy that is appropriate for the subject matter and the audience. Tailoring a strategy is often a trial-and-error process. You envision a path, you try it, and then you look back to see whether it works for your subject matter and audience.

Creating Sections and Subsections. For scientific documents that are longer than a couple of pages, having sections and subsections becomes important. Why? One reason is that sections and subsections show readers the strategy of the document. The headings and subheadings act as a roadmap for readers. When the headings and subheadings are well-written, the readers can quickly see the document's organization. Sections and subsections also provide readers with white space. Readers of scientific papers and reports need white space so that they have time to rest and reflect on what they have read. Besides showing strategy and providing white space, sections and subsections allow readers to jump to information that interests them. Along the same lines, sections and subsections allow readers to skip information that does not interest them. Remember: The primary purpose of your writing is not to entice readers into reading every word you've written, but to inform or persuade your audience as efficiently as possible.

How long should your sections and subsections be? As with most questions about style, there is no absolute answer. If your sections are too long to read in one sitting, your readers will tire in the same way that a driver tires from a long stretch of highway. On the other hand, if your sections are too short, your paper or report will appear as a collage of titles and subtitles. The unnecessary white space will cause your readers to make too many starts and stops. The overall effect is that your readers will tire much in the same way that a driver tires from the starts and stops of congested city traffic.

How should you title a section? When creating titles for sections, you should strive for the same clarity and precision that you have attained in the title of a document. Don't resign yourself as many scientists and engineers do to cryptic one-word titles that clue the readers to nothing:

Slurry
    Combustion
    Pollution
Dry
    Combustion
    Pollution

These titles are vague. Because readers often skim through documents to look for particular results, you want your heading titles to indicate the sections where those results can be found.

Coal-Water Slurry
    Combustion Efficiency
    Combustion Emissions
Dry Pulverized Coal
    Combustion Efficiency
    Combustion Emissions

When creating titles for sections, you should also consider the parallelism of the titles. In other words, don't write

Mining the Coal
Transportation Stage
Burning the Coal

The second heading is not parallel to the other two. Think of your sections as pieces of a pie. It makes no sense to slice a pie and have one piece be apple and another be pecan. If your first subsection title is a noun phrase, then all the subsection titles of that section should be noun phrases. Likewise, if your first subsection title is a participial phrase, then all the subsection titles of that section should be participial phrases.

Noun Phrase	**Participial Phrase**
Mining Stage	Mining the Coal
Transportation Stage	Transporting the Coal
Combustion Stage	Burning the Coal

Finally, if you break your information into one subsection, you must have a second. Having a single subsection is similar to slicing a pie and ending up with only one piece:

Precombustion Processes
    Coal Cleaning
Combustion Processes
Postcombustion Processes

Because "Coal Cleaning" has nothing to be parallel to, this breakdown is inherently non-parallel. You should either include another subsection beneath "Precombustion Processes," such as "Coal Switching," or drop the "Coal Cleaning" subsection:

Precombustion Processes	Precombustion Processes
Coal Cleaning	Combustion Processes
Coal Switching	Postcombustion Processes
Combustion Processes	
Postcombustion Processes	

A good test for headings is how well they reveal the organization of the document when they stand alone as a table of contents. If they do not reveal the organization, then you should reconsider them. In the following example from a progress report on the forensic investigation of Pan Am Flight 103, the weak headings on the left suffer from a number of problems: vague descriptions, non-parallelisms, and single-item subheadings.

Weak Headings	**Strong Headings**
Debris Recovered	Completed Work
Cataloguing	Recovering Debris
Interpretation	Cataloguing Debris
Results	Interpreting the Debris
Placement	Preliminary Results of Work
Bomb Makeup	Placement of Bomb
Work to Be Done	Construction of Bomb
Interpretation	Future Work

In the revision on the right, notice the parallelism on the heading level and in each subheading grouping. Also notice that the revision reveals the overall strategy for the document.

Endings of Documents

The ending of a scientific document provides closure. The ending contains the conclusion sections (of the main text) as well as the back matter. Just as readers have certain expectations for an introduction, readers have certain expectations for the conclusion sections. First, in the conclusion sections, readers expect an analysis of the most important results from the document's discussion. Second, readers expect a future perspective on the work. While readers have certain expectations for the conclusion sections, they generally do not for the back matter. Because of that, back matter sections vary considerably. In a document such as a journal article, the back matter is usually nothing more than a list of references. In a formal report, though, the back matter might contain several appendices, a glossary, an index, as well as a bibliography.

Writing Conclusion Sections. Readers have two expectations for the conclusion sections of a scientific document: an analysis of the key results from the document's middle and a future perspective on the work. What's the difference between the analysis in the conclusion sections and the analysis in the discussion? In the conclusion's analysis, you treat the results as a whole, rather than individually as you did in the discussion. Note that in this analysis you should not act like Perry Mason and bring in new evidence that unravels the mystery of your project. In other words, the conclusion's analysis should arise from the findings presented in the discussion.

Besides presenting an analysis of the key results in the conclusion sections, you also give a future perspective on the work. In some documents that future perspective might be recommendations. In other documents that future perspective might be a nod to the direction in which your research will head. A third kind of future perspective is to mirror the scope and limitations that you presented in the beginning of the document. In the document's beginning, you started with a "big picture" and focused until you reached the scope and limitations of the work. In the conclusion, you now take the work's results that you discussed in the document's middle and show the ramifications of those results on the big picture. In a sense, you complete a circle because in the document's beginning, you started with the big picture, and here you end with the big picture.

How long should a conclusion be? For a short paper, a conclusion may be only one paragraph or even one sentence:

> These tests showed that a nonpowered igniter for lean hydrogen-air mixtures is feasible, and that such an igniter could contribute to the safety of light water nuclear reactors by igniting safe concentrations of hydrogen during a loss-of-coolant accident [Thorne and others, 1985].

Typically, a conclusion section runs the length of an informative summary—about 5 to 10 percent of the length of the main text. What's the difference between a conclusion and an informative summary? Sometimes, very little. However, a conclusion addresses an audience that has read the document, while an informative summary does not. Because of this difference in audience, a conclusion gives you the chance to go into more depth on the results and recommendations.

Another way to look at a conclusion is to see it as bringing together the loose ends of your work. Although you typically cannot tie everything into a neat package, you can convey some sense of closure to your audience. In other words, you don't have to reach a summit in your conclusion, but you should arrive at a plateau.

Consider a conclusion section to a report [Jansen, 1991] about the forensic investigation into the downing of Pan Am Flight 103. Notice that the future perspective in this paper is a series of recommendations.

Conclusions

> The bombing of Pan Am Flight 103 on December 21, 1988, was the worst aviation accident in British history. All 259 passengers on board the aircraft as well as 11 residents of Lockerbie, Scotland, where the plane fell, were killed. The

mid-air explosion spread wreckage from the plane over a 1000 square mile area, and more than one thousand workers were needed to collect the debris of the plane [Brown, 1989].

Summary of Investigation. The forensic investigation into this disaster combined scientifically advanced techniques to reach certain conclusions about the responsible terrorists and their methods. When the bomb aboard Flight 103 exploded, it sent thousands of pieces of debris to the ground. The main way that authorities recovered this debris was by a ground search. The search party consisted of over one thousand volunteers, police, and soldiers, and the search area was about 1000 square miles. This area the authorities divided into twelve sections—each section being about 80 square miles. In addition to the ground search, authorities used infrared photography from satellites and low-flying airplanes. As the wreckage was collected from the ground around Lockerbie, it was brought to an empty airplane hangar several miles from the crash site, where technicians slowly reconstructed the Boeing 747.

In the investigation, one of the first questions that officials asked was, "Where was the bomb on the plane?" During the explosion, the temperatures and pressures inside the cargo hold of the plane reached enormous levels. Temperature and pressures of this magnitude cause certain changes in metals. By studying these changes, forensic analysts estimated the bomb's location. After inspecting many pieces of the plane, analysts concluded that the bomb had exploded just under the "P" in the Pan Am logo. Analysts also concluded which cargo bay (14L) had stored the bomb.

A gas chromatograph analyzed the chemical composition of each piece of debris that was brought in from the crash site. The chromatograph told researchers how much residue from the bomb was on the piece of debris. From the chromatograph findings, researchers decided that the bomb had been located in a copper-colored Samsonite suitcase [Emerson and Duffy, 1990].

Researchers also determined that the bomb placed aboard Flight 103 was technologically advanced. By comparing the chemical composition of the bomb residue to compositions of known explosives, researchers concluded that the explosive was Semtex, a Czech-made plastic explosive. From the tiny fragments of the bomb imbedded in the items around it, bomb experts learned that the bomb used a two-step detonator, which exploded the bomb only after both detonators were activated. The first detonator was a barometric detonator that went off when the plane's altitude caused the pressure inside the cargo bay to dip below a certain level. The second detonator was a simple timer.

One of the parts of the bomb that authorities recovered was a microchip from the detonator circuit. This microchip had the same structure as a microchip found on two Libyan agents who in 1986 were caught carrying twenty pounds of Semtex into Senegal [Wright and Ostrow, 1991].

Something that confused investigators in the early part of the investigation was that the pilot had not sent any distress signal. Although the plane began to break up soon after the bomb detonated, authorities felt that the pilot would have had time to send a "Mayday" call. However, once it was concluded that the bomb had detonated in cargo bay 14L, airplane experts realized that the bomb had damaged the plane's electronics center. This center receives electrical energy from the plane's engines and distributes it to every electronic device on the plane. When the bomb damaged this electrical station, the radios used to send distress signals became useless [Emerson and Duffy, 1990].

From the collected pieces of plane wreckage, experts were able to tell how the plane disintegrated. The explosion produced one large hole in the fuselage and another in the main cabin floor of the forward cargo hold. The pressure blast of the bomb caused large cracks to develop along the fuselage and floor, even though the aircraft had been specially strengthened to carry military freight during national emergencies. The cockpit, nose, and forward cabin then separated from the rear section of the plane [Shifrin, 1990].

Perhaps the most important result of the investigation was that authorities collected enough evidence to bring the case to trial. The microchip recovered from the bomb's detonator linked the regime of Libya's Moammar Gadhafi to the bombing. Authorities believe that the bombing was in retaliation for the 1986 bombing of Tripoli by the United States [Wright and Ostrow, 1991]. Also, the cargo bay containing the bomb held many bags from Malta, a country closely allied with Libya. Although two Libyans have been identified as being responsible for planting the bomb, authorities still have not been able to extradite them.

Recommendations of Investigation. The investigation into the bombing of Pan Am Flight 103 led to several recommendations to help prevent another explosion of this kind. Some of these recommendations were for changes to airplane construction; other recommendations were for changes in airport safety.

Recommendations for changes in airplane construction fell into two categories: changes in the cargo-bay design and changes in the flight-recorder apparatus. After reviewing the investigation, the Air Accidents Investigation Branch of the British Transport Department recommended that all luggage be contained in stronger cargo bays. Although authorities admit that such measures could not have prevented the Flight 103 disaster, they feel that stronger cargo bays could allow planes to survive explosions of smaller bombs [Shifrin, 1990].

Other suggestions for changes to airplanes concerned the flight recorders. Because the bomb of Flight 103 cut power to the flight recorders, the recorders were of no help to the investigation. Part of the problem was that the voice recorders had no power backup. Furthermore, several minutes of recordings were stored in volatile memory (which is erased in a power failure) before being transferred to magnetic tape. Therefore, not only were investigators unable to hear what happened in the plane after the power went out, but they were also unable to hear what happened just before that time. The Air Accidents Investigation Branch recommended that flight recorders have back-up batteries, and that their volatile memory be replaced with non-volatile memory [Shifrin, 1990].

Although the proposed changes to airplanes would certainly help reduce the effects of bombs and make the subsequent investigations easier to carry out, keeping bombs off planes was made a higher priority. To do so, authorities imposed several new security restrictions on airports, particularly those in Europe and the Middle East. First, authorities insisted that each bag correspond to a passenger, and that if a passenger gets off the plane, the corresponding bags get off as well. Second, authorities began randomly searching passengers and their bags. Last, authorities stepped up plans to install sophisticated devices capable of detecting plastic explosives such as Semtex [Watson and others, 1989].

Writing the Back Matter. Rarely will you write a report for only one type of audience. Most scientific reports have several types of readers, each type with a different technical background and reason for reading the report. How then do you write the main text of your report for all these audiences? The answer is that you don't. You write the text of your report for your main audience, and you supply back matter in the form of appendices and glossaries for your secondary audiences.

Often, you write appendices to give additional information to secondary audiences. This information can take many forms. For instance, a common type of appendix presents background information to a less technical audience. For example, if you had written a report on improving a chemical test for the forensic analysis of blood, you might include for less technical readers an appendix [Mickey, 1993] explaining the analysis of bloodstains. As with any appendix, this appendix should stand on its own as a separate document with a beginning, middle, and ending.

Appendix
Analysis of Bloodstains

Forensic serology is an important field in forensic science because bloodstains are frequently obtained at crime scenes involving homicides, rapes, and assaults. During an examination of a suspected stain, the forensic serologist must answer three questions:

(1) Is it blood?
(2) If it is blood, is it human?
(3) If it is human blood, how closely can it be associated to a particular individual?

To answer these questions, the forensic serologist performs several tests on the stains [Saferstein, 1981].

Two blood identification tests are the phenolphthalein test and the luminol test. The phenolphthalein test is a catalytic color test that produces a deep pink color when blood, phenolphthalein, and hydrogen peroxide are mixed. The luminol test, unlike the phenolphthalein test, results in the production of light rather than color. The luminol test is used exclusively by investigators to detect small traces of blood and unusual bloodstain patterns [Lee, 1982].

After identifying a stain as blood, the serologist determines whether the bloodstain is of human origin. The precipitin test is the standard test used in forensics to determine the species of a bloodstain. By injecting an animal (usually a rabbit) with human blood, antibodies will form in the animal that react specifically with the human blood. The animal is then bled, and the blood serum is isolated. The blood evidence is layered on top of the serum in a test tube. If the blood evidence is human, a white band or cloudy ring will form at the interface of the two liquids.

The last and most important step in analyzing a bloodstain is to associate the blood to a particular individual. The traditional methods (all blood analysis methods prior to DNA fingerprinting) of tagging a bloodstain to a person require the serologist to determine the combination of blood factors in the blood sample. If a sufficient number of the blood factors can be determined, the probability of an individual having that combination of blood factors is determined by taking the product of each blood factor's frequency in the population. The traditional methods are accurate for a fresh blood sample, but most blood evidence is received in the form of dried blood stains. Few of the blood factors survive the drying and aging of a bloodstain.

DNA fingerprinting is the most accurate test used in forensics to tag a bloodstain to an individual. An advantage of DNA fingerprinting is that DNA molecules can be detected in dried blood. The most common DNA fingerprinting technique is called "restriction fragment length polymorphism" analysis, or RFLP analysis. In this analysis, RFLP patterns are visible after transferring the DNA fragments to an x-ray film. The RFLP pattern, which is similar to a bar code on groceries, is the final product of a DNA fingerprint. When the bars of two samples match, scientists conclude that the samples came from the same person. False identification of a suspect is avoided with DNA fingerprinting because degraded DNA will not produce a different RFLP pattern. As the DNA degrades, the overall RFLP pattern becomes weaker, but individual RFLP patterns are neither created nor destroyed [McNally, 1989].

Forensic scientists have studied the environmental effects of the integrity of DNA samples. These environmental effects include heat, humidity, soil, and ultraviolet light. The results of these experiments have shown that only soil contamination affects the integrity of DNA isolated from bloodstains. However, the integrity of the DNA is not altered such that false patterns are obtained. This finding substantiates the claim that DNA will not identify the wrong suspect [Shaler, 1989].

Forensic experts envision a national computer file of the DNA types of convicted felons. The California legislature already requires that sex offenders and murderers submit a sample of their DNA upon release from prison. A DNA database, similar to the FBI's fingerprint database, could revolutionize law enforcement.

Another type of appendix is one having detailed information for a more technical audience. For instance, in a report on the forensic use of gas chromatography and mass spectrometry, you might include an appendix for more technical readers on the types of mass spectrometers. This appendix could explain the workings of four common types of mass spectrometers: time-of-flight, magnetic sector, quadrupole, and ion trap. In this appendix, you could provide diagrams to explain how each spectrometer works.

Still a third kind of appendix is one that presents branch information. Sometimes you want to include secondary information that is interesting, but not directly pertinent to the results you're emphasizing. For a report on forensic techniques, that secondary type of information could be a case study, such as the case of the Birmingham Six. In this case, six men were wrongly convicted of bombing two pubs in Birmingham, England. The men were convicted on the basis of a single test, called the Greiss Test, which detected amounts of nitroglycerine. The men tested positive. Years later, while researching the Greiss test, scientists discovered that contact with many substances such as playing cards, adhesive tape, and plastic wrappers from cigarette packages produced false positives [Hamer, 1991]. Although information about this case of the Birmingham Six is not necessary for understanding your work, you still could include the information for the sake of completeness or audience interest. In such a situation, an appendix would be an appropriate place for the information.

For smooth transitions in the document, there should be at least one mention of each appendix somewhere in the main text of the report. In other words, when the occasion arises in the text, refer readers to the appendix. For example,

> The mass spectrometer must also be in a high vacuum to minimize the number of gas molecules that collide with the ions. For more information on the different types of mass spectrometers, see Appendix B.

A glossary is a special appendix that gives background definitions to secondary audiences. Let's say the primary audience for a report on the effects of spaceflight on the human immune system were immunologists and the secondary audience included NASA management. In the back of the report you might include the glossary [Bodden, 1993] given below. This glossary allows you to inform the secondary readers about the vocabulary of the report without breaking the continuity of the writing for the primary readers. Notice that if the primary readers had been NASA management, it would have been appropriate to define these terms in the text.

Glossary

antibody: a protein molecule that is released by a daughter cell of an activated B cell. Antibodies bind with antigens and serve as markers that give signals to immune cells capable of destroying the antigens.

antigen: a substance or part of a substance that is recognized as foreign by the immune system, activates the immune system, and reacts with immune cells and their products.

cytoxic: a type of activity related to destructive capabilities. Cytoxic can be used interchangeably with the word "killing."

humoral: of or pertaining to body fluids.

immune response: a defensive response by the immune system as a reaction to detection of an antigen. T cells and B cells detect antigens after the macrophage has signaled that an antigen is present. This detection by the T cells and B cells provokes the cells to respond; thus, they become activated.

immunocompetent: ability of the body's immune cells to recognize specific antigens. When T cells and B cells become immunocompetent, they are able to attack antigens.

immunodeficiency: a disease resulting from the deficient production or function of immune cells required for normal immunity.

killer T cell: a type of T cell that directly kills foreign cells, cancer cells, or virus-infected body cells.

white blood cell: a type of body cell that is involved in body protection and takes part in the immune response. For instance, lymphocytes are a specific type of white blood cell.

In creating a glossary, arrange terms in alphabetical order. Use italics or boldface in the text to key readers to the terms that the glossary will define. As with appendices, the glossary should have a direct connection to the text of the document.

Note that some documents have box stories, also called "sidebars," that fill the same role as appendices and glossaries in a report. Instead of falling at the end of the document, these box stories are formatted alongside the text so that a secondary audience can stop and read them. In practice, though, some documents have so many box stories that primary audiences need a map to find out which paragraphs in the main text to read next. In such cases, the writer has sacrificed the primary audience's continuity in reading for the chance to give background or detour information to the secondary audience—not a good trade.

Hypertext, which is a form of writing for computer documents such as those to be found on the World Wide

Web, overcomes this problem by placing box stories in hidden computer windows. The readers then have a choice: continue reading the text or access the window by clicking on a color-coded word. In essence, what hypertext provides is an efficient way for secondary readers to reach the back matter without interrupting the main text for the primary readers.

References

Bodden, C., "Effects of a Long-Duration Space Mission on the Human Immune System," *Undergraduate Engineering Review* (1993), p. 39.

Brown, David A., "Investigators Expand Search for Debris From Bombed 747," *Aviation Week and Space Technology*, vol. 130, no. 25 (January 9, 1989), pp. 26–27.

Bullard, Fred M., *Volcanoes of the Earth*, 2nd ed. (Austin: University of Texas Press, 1976).

Emerson, Steven, and Brian Duffy, *The Fall of Pan Am Flight 103* (New York: G. P. Putnam's Sons, 1990).

Garberson, Jeff, "Climate Change: From the 'Greenhouse Effect' to 'Nuclear Winter,'" *The Quarterly* (Livermore, CA: Lawrence Livermore National Laboratory, July 1985), p. 98.

Hamer, M., "Forensic Science Goes on Trial," *New Scientist* (November 9, 1991), pp. 30-33.

Janzen, P., "The Investigation Into the Fall of Pan Am Flight 103," *Undergraduate Engineering Review* (1991), pp. 53–58.

Kelsey, James, "Inertial Navigation Techniques Improve Well-Bore Survey Accuracy Ten-Fold," *Sandia Technology*, vol. 7, no. 1 (1983), p. 22.

Kunhardt, Erich, and W. Byszewski, "Development of Overvoltage Breakdown at High Pressure," *Physical Review A*, vol. 21, no. 6 (1980), p. 2069.

Lawrence Livermore National Laboratory (LLNL), "Exotic Forms of Ice in the Moons of Jupiter," *Energy and Technology Review* (Livermore, CA: Lawrence Livermore National Laboratory, July 1985), p. 98.

Lee, H. C., "Identification and Grouping of Bloodstains," *Forensic Science Handbook*, 2nd ed., ed. by R. Saferstein (Englewood Cliffs, NJ: Prentice-Hall, 1982), pp. 267–327.

McNally, L., "The Effects of Environment and Substrata on Deoxyribonucleic Acid (DNA): the Use of Casework Samples From New York City," *Journal of Forensic Sciences*, vol. 34, no. 5 (September 1989), pp. 1070–1077.

Mickey, C., "Analysis of Bloodstains," *Undergraduate Engineering Review* (1993), p. 12.

Perry, R. A., and D. L. Siebers, "A New Method to Reduce Sulfur Dioxide Emissions From Diesel Engines," *Sandia Combustion Research Program Annual Report* (Livermore, CA: Sandia National Laboratories, 1985), chap. 3, p. 3.

Saferstein, R., *Forensic Science Handbook* (Englewood Cliffs, NJ: Prentice-Hall, 1982).

Shaler, Robert, "Evaluation of Deoxyribonucleic Acid (DNA) Isolated From Human Bloodstains Exposed to Ultraviolet Light, Heat, Humidity, and Soil Contamination," *Journal of Forensic Sciences*, vol. 34, no. 5 (September 1989), pp. 1059–1069.

Shifrin, Carole A., "British Issue Report on Flight 103, Urge Study on Reducing Effects of Explosions," *Aviation Week and Space Technology*, vol. 133, no. 12 (September 7, 1990), pp. 128–129.

VanDevender, P., "Ion-Beam Focusing: A Step Toward Fusion," *Sandia Technology*, vol. 9, no. 4 (December 1985), pp. 2–13.

Wright, Robin, and Ronald J. Ostrow, "Libya Regime Suspected in Pan Am Blast," *Austin American Statesman* (June 24, 1991), section A, pp. 1, 4.

Chapter 3

Structure: Providing Transition, Depth, and Emphasis

Science is built up with facts, as a house is with stones. But a collection of facts is no more science than a heap of stones is a house.

—J. H. Poincaré

Structure is not just the organization of details. Although organizing details in a document is certainly important, many well-organized documents fail to inform because the writer has not made strong transitions between the details or has not presented the details at the proper depth or has not placed the proper emphasis on the details.

Transitions Between Details

In a scientific document, you make transitions not only between sentences and paragraphs, but between sections. You may organize your paper into logical sections, but if you don't make transitions between those sections, you can lose your readers.

In the previous chapter, several reasons were given

for the use of sections in scientific writing: to help emphasize details, to reveal the logical strategy of the details, and to parcel out details so that the information is easier to digest. A disadvantage of using sections is that they make inherent discontinuities in the reading. In other words, between each section there is white space, which makes the readers stop. Two ways to overcome these discontinuities between sections are to map the sections and to smooth the entrances into the sections.

Mapping sections is straightforward. When you divide a discussion into sections, you seek divisions that are logical and appropriate for the audience. Perhaps the section names form a sequence of steps, such as the stages of combustion, or perhaps the sections are parallel parts and sum to a whole (the three parts of a comet—head, coma, tail).

How do you map sections? The simplest way is to present the names of the sections in a list just before the sections occur. Consider an example [Saathoff, 1991] that maps three subsections:

> **Dangers of Breathing Compressed Air**
>
> Recreational scuba divers breathe compressed air at depths down to 190 feet. Breathing compressed air at these great depths and even at more moderate depths poses many dangers for scuba divers. These dangers are nitrogen narcosis, decompression sickness, and arterial gas embolism.

Now the readers are prepared for the three subsections that follow: nitrogen narcosis, decompression sickness, and arterial gas embolism.

Besides using mapping to show readers the divisions in a document, you also make transitions by smoothing the entrances into sections. Because readers often skip around in papers and reports, you should make each section and subsection stand on its own. How should you begin a section or subsection? Perhaps an easier question to answer is how shouldn't you begin.

There are three common beginnings to a section that you should avoid. One is the "empty" beginning: Because the first sentence of a section receives heavy emphasis, you want to say something in that first sentence. In an empty beginning, the writer wastes the first sentence and loses an opportunity to inform:

> The behavior of materials in combustion systems has considerable technological significance.

In this sentence, nothing occurs. In other words, the scientist did not advance the knowledge of the audience. The scientist also failed to build a foundation for the rest of the section.

The second type of beginning to avoid is the *"in medias res"* beginning. Here, the writer begins with details that are too specific:

> Item 12 on Drawing XLC-3549 shows the tolerance of the safe adapter.

In this beginning, the engineer quickly narrowed the audience until the only person following the discussion was the author himself.

The third type of beginning to avoid is the "Genesis" beginning. While the *in medias res* beginning opens with details that are too specific, the Genesis beginning opens with details that are too general:

> Man has since the beginning of time attempted to acquire a greater and greater control over his environment. Gaining control over a situation serves not only a survival-related need, but also a psychological need. Man's need for better control of his environment has increased greatly during and following any time of major conflict, such as World War II. This need and desire for control is evident in all technological settings, including the welding field.

By conjuring images of prehistoric man and the Second World War, the engineer has created expectations that a section on welding, no matter how well written, just can't deliver.

So far, we've discussed ways not to begin a section. What then are some proper ways? One straightforward way is to introduce the subject of the section:

> ***Precombustion Processes.*** Precombustion processes for reducing emissions of sulfur dioxide occur before the coal is even burned. There are two major types of precombustion processes: coal switching and coal cleaning.

In this beginning, you include the section heading in the first sentence. Don't think that beginning a section with this repetition of the heading is redundant. A redundancy is a needless repetition. The repetition of the heading in the first sentence of the section is an important repetition because it reinforces the identity of the section. In a sense, this repetition strengthens the audience's confidence in you to deliver what you are promising to deliver. Contrast that straightforward repetition with an abrupt beginning:

> ***Precombustion Processes.*** These occur before the combustion process. There are two major types...

This abrupt beginning throws readers off-balance. The readers don't immediately gather that the pronoun "these" refers to precombustion processes.

The straightforward repetition of the heading is the most common way to begin a section or subsection. Sometimes, though, a straightforward beginning is not the most efficient way to dive into a section or subsection. Consider a situation in which your heading contains a term that your audience doesn't understand. In this situation, you may want to begin with background information and then focus on the subject of the section.

> ***Downhole Steam Generator.*** More than half of the oil in a reservoir is too viscous to pump out with conventional methods. By heating these oils with steam and decreasing their viscosity, we can recover billions of gallons. For oils below 800 meters, the steam produced on the surface loses too much energy in transit to heat the oil. One way to overcome this

problem is to use a downhole steam generator that applies hot steam directly.

In this example, the title of the subsection was a term, "downhole steam generator," that was unfamiliar to the audience. Instead of beginning the subsection with a repetition of that term, the writer chose to give background information that would allow the audience to understand the term. There are two important criteria for using this strategy. First, the background information should engage the audience. Second, the background information should be brief; it should quickly focus on the term contained in the subheading. If your background is long-winded, you will tax your readers' patience.

The following section uses all the techniques discussed so far to make transitions [Smith, 1993]. In the first paragraph, the writer begins with background information before quickly focusing on the section's topic—avoiding high altitude illness. This first paragraph also contains a mapping sentence for the subsections. Finally, each subsection begins with a straightforward sentence that repeats the subheading.

Avoiding High Altitude Illness

The sport of mountain climbing can involve dangers such as hanging from a precipice thousands of feet above the ground and weathering extremely cold temperatures; however, the simple act of breathing while at high altitude introduces a new class of dangers to the sport. High altitude illness can temporarily debilitate a climber and even lead to the climber's death. There are three common high altitude illnesses: acute mountain sickness, high altitude pulmonary edema, and high altitude cerebral edema. Knowledge of these three commonly reported illnesses can help prevent disaster in a mountain climbing expedition.

Acute Mountain Sickness. Acute mountain sickness is the mildest of the three altitude illnesses. A case of this illness usually lasts four or five days after arrival at high altitude [Ward and others, 1989]. Victims of acute mountain sickness have symptoms such as headaches, listlessness, fatigue, and drowsiness [Foster, 1983]. Doctors are not sure of the exact cause of this illness, although they believe that oxygen deprivation due to altitude is the starting mechanism [Heath and Williams, 1979].

Treatment of acute mountain sickness involves helping the victim adjust to high altitude [Ward and others, 1989]. Once the victim adapts to high altitude, the symptoms disappear. To prevent this illness, doctors recommend a slow and steady climbing pace [Foster, 1983]. Other preventive measures are to eat foods high in carbohydrates and to drink large amounts of liquids.

High Altitude Pulmonary Edema. High altitude pulmonary edema is an altitude illness in which fluid fills the lungs. Victims of this illness have usually ascended quickly to a high altitude and engaged in strenuous physical activity immediately after arrival. High altitude pulmonary edema occurs less often than the other illnesses; an incidence rate of only 2.5 percent was reported by a Mount Everest check station that stands at a 4,243 meter altitude [Ward and others, 1989]. Even though the incidence rate is relatively low, this illness can kill its victims in hours, warranting the attention of mountain climbers. Symptoms of high altitude pulmonary edema include the symptoms of acute mountain sickness along with cardiopulmonary symptoms such as cough, chest pain, and heart complaints [Lobenhoffer and others, 1982]. The causes of high altitude pulmonary edema are unknown, but doctors speculate that high pressures in the blood vessels of the lungs force fluid into the airways [Ward and others, 1989].

Decreasing altitude is the best method of treating this illness [Heath and Williams, 1979]. Medicines that force the body to expel fluid and combat infection are also used to treat this illness. Another treatment used by doctors involves breathing techniques that increase the pressure inside the lungs. Finally, preventing high altitude pulmonary edema involves three key measures: slow ascent, limited exertion at altitude, and abstinence from alcohol. Each of these measures minimizes stress on the cardiovascular system.

High Altitude Cerebral Edema. The third high altitude illness is high altitude cerebral edema. In this illness, the brain swells with fluid, creating pressure on the skull [Heath and Williams, 1979]. Damage to the brain from high pressure can be permanent, making high altitude cerebral edema the most

> serious of the three illnesses discussed here. Victims of this illness show signs of acute mountain sickness early, as in the case of high altitude pulmonary edema [Ward and others, 1989]. After time, the victims become irritable and irrational, have hallucinations, and slip into unconsciousness.
>
> The primary treatment of high altitude cerebral edema is, as with high altitude pulmonary edema, moving the victim to a lower altitude. Treatments also include medications administered to remove excess fluid from the brain. Similar to high altitude pulmonary edema, high altitude cerebral edema can be prevented by keeping a slow rate of ascent, and limiting activity after arrival at high altitudes.

Notice that each of the three subsections followed the same sequence of topics: description of illness, symptoms of illness, treatment of illness, and prevention of illness. Having the same sequence in each subsection helps smooth the transitions within the subsections because after the first subsection, the readers expect these details to come in a certain order.

Depth of Details

Depth is the level of detail that you, the writer, provide for a subject. Depth goes much deeper than just the number of details that you accumulate. Depth also includes the way in which you classify, analyze, and assess those details. For instance, if you reviewed the literature on predictions for how global warming will affect sea levels, you could write a paper that would simply classify the predictions done by others. Here you would state who predicted what and place each prediction in an appropriate category. A deeper review would go beyond classification to include an analysis of the predictions. For instance, in this deeper review, you could, as Houghton and Woodwell [1989] did, use predictions to calculate how an anticipated rise of 4–5 meters would affect the coastline of Florida. In the analysis, Houghton and Woodwell showed that most coastal cities, including Miami, Fort Myers, Daytona Beach, and St. Petersburg, would be submerged. In an even deeper review, you would not just analyze the measurements, but also assess the validity of the predictions. Here, you would establish criteria for deciding if or when each prediction was valid and then evaluate the predictions based on the criteria.

Choosing a proper level of detail is difficult because depth does not depend on just one variable. First, depth depends on format. Format determines depth in a simple way. If your format limits the number of words, then your depth is accordingly limited. For instance, if you have only one paragraph to discuss the problem of chlorofluorocarbon pollution, the depth might be as follows:

> Chlorofluorocarbons are man-made chemicals that are commonly used in water chillers, air conditioners, and aerosols in the refrigeration and electronics industries. Chlorofluorocarbons destroy the ozone layer and also contribute to the greenhouse effect. To reduce chlorofluorocarbon emissions, scientists are exploring three strategies: (1) substitutes for chlorofluorocarbons, (2) system designs that recycle chlorofluorocarbons, and (3) system designs that don't need chlorofluorocarbons.

In this example, the chemist brings out three related points: a definition of chlorofluorocarbons, the effect that chlorofluorocarbons have on the environment, and industry's response to these regulations.

Notice how the depth attributed to each point changes if the chemist has a page, rather than a paragraph, to communicate the information.

> Chlorofluorocarbons (CFCs) are man-made chemicals that are commonly used in water chillers, air conditioners, and aerosols in the refrigeration and electronics industries. CFCs are chemically similar to hydrocarbon atoms, except that hydrogen atoms are replaced by chlorine, fluorine, or bromine atoms. This substitution makes the CFCs chemically stable in the lower atmosphere. However, when these

molecules drift to the upper atmosphere, they are broken down by ultraviolet light from the sun.

In the upper atmosphere, there exists an ozone layer that serves as a protective shield against the sun's ultraviolet radiation [Downing, 1988]. Without this shield, harmful levels of ultraviolet radiation would strike the earth's surface causing injury and possible death to people, animals, and plants [Reitz, 1990]. When a chlorofluorocarbon molecule breaks down in the upper atmosphere, the free chlorine atom can begin a chain reaction that can destroy tens of thousands of ozone molecules. During the past decade, the ozone layer has shrunk by about 2.5 percent. Scientists have attributed this reduction to the presence of chlorofluorocarbons, which have been in use for the past fifty years [Reinhardt and Groeneveld, 1989].

Because manufacturers weren't aware of the ozone depletion problem until only the last few years, they have not had time to produce many alternatives. After development of an alternative, about five to seven years of testing are needed before the chemical can be marketed. For that reason, most of the existing CFC substitutes are not available commercially. Equally important, many alternatives are not fully compatible with existing equipment.

Here, instead of one paragraph with three points, we have three full paragraphs. Notice that if the format allowed for an entire article on the subject, the writer could expand further and write a section instead of a paragraph for each point. In this further revision, the first section that gives the definition of chlorofluorocarbons could include illustrations and chemical equations.

The audience determines the level of detail in a more complicated fashion. First, the interest of the audience in the subject affects the depth. If the reader desires much depth about the topic, you are expected to provide it. However, as you present details about a subject, you also spark questions about the subject. For that reason, achieving a proper depth means finding a level such that you satisfy the reader's interest and anticipate the reader's questions. Consider the following three depths for the same topic:

Depth 1: The Environmental Protection Agency has tightened emission standards by 60 percent.

Depth 2: The Environmental Protection Agency has tightened emission standards from 0.25 g/hp-h to 0.1 g/hp-h.

Depth 3: The Environmental Protection Agency has tightened emission standards from 0.25 g/hp-h (grams of particulate/horsepower-hour) to 0.1 g/hp-h. The particulates include hydrocarbons, carbon monoxide, and nitrogen oxides. All three particulates are considered pollutants.

Of these three depths, the first is the most shallow because it gives the least amount of information, and the third depth is the deepest because it gives the most information. Although the second depth is deeper than the first depth, it is less successful because it raises unanswered questions—namely, what does the abbreviation "g/hp-h" mean? In a sense, this depth is simultaneously too shallow and too deep. Once you raise questions at a certain depth, you are obliged to answer those questions.

The technical level of the audience is a second way in which audience affects depth. The more technical the audience, the more quickly you can achieve a technical depth because with a technical audience, you need not provide as much background as you do with a non-technical audience.

A third way in which audience determines depth has to do with the purpose of the document. For instance, in an informative paper about how effective photorefractive keratectomy has been at correcting nearsightedness, you present the results of the procedure: what percentage achieved normal vision, what percentage achieved significantly improved vision, what percentage had complications, what percentage experienced regression after a year. However, in a paper arguing for approval of the technique, you go deeper to include rebuttal arguments for the problems that the procedure has. You would also include a discussion about the advantages that photorefractive keratectomy has over alternatives such as radial keratotomy.

One quick check for the depth of a document is to examine the lengths of paragraphs and sections. In general, the longer the paragraphs and sections, the greater the depth that is achieved. When paragraphs and sections are short, the initial impression given to the audience is that the document lacks depth. In some cases, though, the audience doesn't expect much depth. For example, in a set of instructions the audience is primarily concerned with the surface question "how" rather than the deeper question "why." For that reason, instructions often have short sections and short paragraphs that answer the question of "how," but do not engage the question of "why." In contrast, journal articles and formal reports confront both questions. Correspondingly, the sections and paragraphs of those documents are longer.

Can sections and paragraphs be too long? In scientific writing, the answer is yes. Although a fiction writer such as William Faulkner can pull off a four-page paragraph, a scientific writer cannot. A scientific writer has to divide the information into digestible portions separated by white space. Otherwise the reader tires, and the efficiency of understanding goes down.

Note that once you establish a depth in discussing a topic, readers expect you to maintain that depth for the discussion of related topics. That balance of depth is particularly important in comparisons, such as the comparison of altitude sicknesses in the previous section.

Emphasis of Details

In any description, a writer will present details that have varying degrees of importance. In some kinds of writing, fiction for instance, emphasis of details is not so important. In fact, a good mystery writer will often bury an important detail. In scientific writing, though, scientists and engineers should present details in such a way that readers understand their relative importance. Emphasizing details is not necessarily difficult; however, many scientists and engineers fail to emphasize details properly in their documents. In fact, lack of emphasis is probably the most common structural problem in scientific writing. How do you go about emphasizing details? To emphasize a detail, you have several means: repetition, wording, illustration, and placement.

Emphasizing Details With Repetition. If you mention a particular result once in your summary, a second time in your discussion, and then a third time in your conclusion, your readers will realize its importance. Repeating important results in a document is not being redundant. You are redundant when you repeat something unimportant. Redundancies, which do not serve a document in any way, usually occur within phrases: "bright green in color," "fundamental basics," or "already existing." In such cases, the phrase could be tightened without loss of meaning: "bright green," "basics," and "existing."

Repeating important details in a document serves a document because it increases the likelihood that readers will retain the important details. Readers don't retain every detail in a document. Studies show that people remember only about 10 to 20 percent of what they read [Felder and Stice, 1992]. Mentioning a detail two or three times in the document helps increase that percentage. Advertisers know the value of repetition. Next time you watch a television commercial, count how many times the commercial repeats the product's name. Don't be surprised if you either hear or see the name five or more times.

The typical organization of most scientific documents gives you the opportunity to use repetition. In the informative summary you have your first opportunity to mention the detail. A second opportunity occurs in the discussion of the work, and a third opportunity occurs in the conclusion. Even in a document with no informative summary, you have at least two opportunities (the discussion and conclusion) for mentioning important details. Those opportunities you should not miss.

Emphasizing Details With Wording. Many details in scientific documents float, ungrounded, because the author has not shown why the details have been included:

> One of the panels on the north side of the solar receiver will be repainted with Solarcept during the February plant outage.

What is the most important detail in the sentence? Is it that the panel is on the north side? Is it that the panel is being repainted with Solarcept? Is it that the repainting will occur during the February plant outage? The problem with this sentence is that you don't know. All details have the same weight. In this example, prepositional phrases are overused. Prepositional phrases list details, but do not give details any relative emphasis. Details linked by prepositional phrases all have the same amount of importance.

In strong scientific writing, the writer anchors the details by giving reasons for their inclusion:

> Because the February plant outage gave us time to repair the north side of the solar receiver, we repainted the panels with Solarcept, a new paint developed to increase absorptivity.

In this revision, readers have an easier time assessing the importance of the details because readers can see why the details are included. While wording without emphasis overuses prepositional phrases, wording with emphasis judiciously uses *dependent clauses* and *infinitive phrases*.

Dependent clauses are clauses that begin with introductory words such as "because," "since," "as," "although," and "when."

> Because the February plant outage gave us time to repair the north side of the receiver,...

Infinitive phrases are *verb* phrases that begin with the word "to."

> ...to repair the north side...to increase absorptivity.

Wording details through dependent clauses and infinitive phrases helps show the relationship and relative importance of details.

Emphasizing Details With Illustration. Another way to accent a detail is through illustration. In a document, readers often don't read every sentence. Although scientific readers don't read every sentence, they almost always look at every illustration. Therefore, if you can place important results in an illustration, do so. For example, Figure 3-1 shows how much radiation the average person in the United States receives from the operation of nuclear power plants as opposed to other sources. These other sources include natural sources (such as radon) and medical sources (such as dental x-rays). Illustrations such as Figure 3-1 stand out in papers and reports. You should realize, though, that a large number of illustrations dilutes the importance given to any one. If Figure 3-1 was one of ten such pie graphs, it would not receive much emphasis.

Emphasizing Details With Placement. Certain places in a document receive more emphasis than others. For instance, text that borders white space receives more emphasis than text that borders other text. For this reason, the titles and headings receive emphasis because they are surrounded by white space (line breaks before and after). Likewise, the beginnings and endings of sections also re-

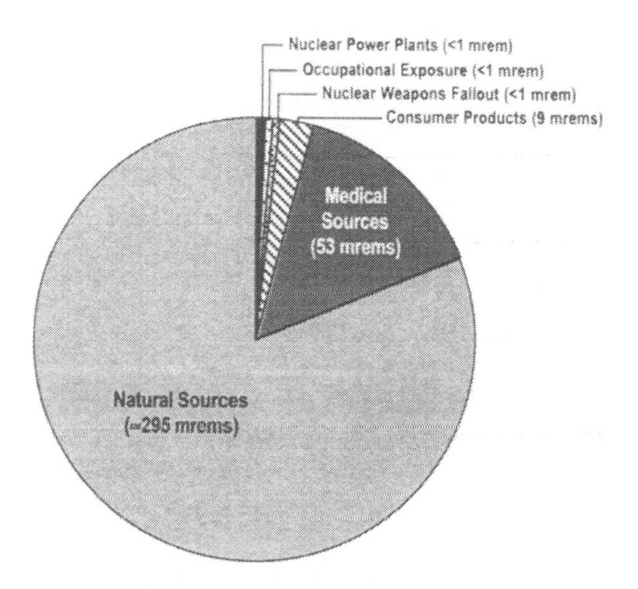

Figure 3-1. The breakdown of annual radiation dosage to the average person in the United States from all sources [*Radiation*, 1994]. Most of the contribution comes from natural sources, such as radon and cosmic radiation.

ceive emphasis because they are bounded, either above or below, by white space.

Other places in documents receive emphasis because of white space, but to a lesser extent. For instance, the beginnings and endings of paragraphs receive emphasis because of the white space given by the tab at the beginning of paragraphs and by the white space at the end of the paragraph's last line.

In addition to using white space for emphasis, you can also use the lengths of sentences and paragraphs. For example, a short sentence following a long sentence receives emphasis, particularly if that short sentence is the last sentence of the paragraph. Likewise, a short paragraph following a long paragraph receives emphasis. In the following example, notice how the Warren Commission [1964] used a combination of short sentences and placement at the end of the second paragraph to emphasize the name of the man in the lunchroom.

When the shots were fired, a Dallas motorcycle patrolman, Marrion L. Baker, was riding in the motorcade at a point several cars behind the President. He had turned right from Main Street onto Houston Street and was about 200 feet south of Elm Street when he heard a shot. Baker, having recently returned from a week of deer hunting, was certain the shots came from a high-powered rifle. He looked up and saw pigeons scattering in the air from their perches on the Texas School Book Depository Building. He raced his motorcycle to the building, dismounted, scanned the area to the west and pushed his way through the spectators toward the entrance. There he encountered Roy Truly, the building superintendent, who offered Baker his help. They entered the building, and ran toward two elevators in the rear. Finding that both elevators were on an upper floor, they dashed up the stairs. Not more than 2 minutes had elapsed since the shooting.

When they reached the second-floor landing on their way up to the top of the building, Patrolman Baker thought he caught a glimpse of someone through the small glass window in the door separating the hall area near the stairs from the small vestibule leading into the lunchroom. Gun in hand, he rushed to the door and saw a man about 20 feet away walking toward the end of the lunchroom. The man was empty-handed. At Baker's command, the man turned and approached him. Truly, who had started up the stairs to the third floor ahead of Baker, returned to see what had delayed the patrolman. Baker asked Truly whether he knew the man in the lunchroom. Truly replied that the man worked in the building, whereupon Baker turned from the man and proceeded, with Truly, up the stairs. The man they encountered had started working in the Texas School Book Depository Building on October 16, 1963. His fellow workers described him as very quiet—a "loner." His name was Lee Harvey Oswald.

Placement can work in the opposite way: Placing important information in the wrong place can greatly reduce the chances that the audience will remember that information. For instance, many scientists and engineers bury important details in the middle of paragraphs.

This report uses data from both the test and evaluation and power production phases to evaluate the performance

> of the Solar One receiver. Receiver performance includes such receiver characteristics as point-in-time steady state efficiency, average efficiency, start-up time, operation time, operations during cloud transients, panel mechanical supports, and tube leaks. Each of these characteristics will be covered in some detail in this report.

Now that you've read this paragraph, close your eyes and name as many receiver characteristics as you can that will be covered in the report. Did you remember all seven items of the list? As stated in the previous chapter, people remember things in groups of twos, threes, and fours. The list here was too long. Also a problem was that the list occurred in the middle of the paragraph. A better way to emphasize this information would be to group those characteristics and then place the list in a location, perhaps the end of the paragraph, that receives more emphasis.

> This report uses data from both the test and production phases to evaluate the performance of the Solar One receiver. In this report, we will evaluate performance by studying the receiver's efficiency, operation cycle, and mechanical wear.

You might ask why not break out of the paragraph form with the list and format it vertically down the page. Because of the additional white space, this vertical list would certainly receive more emphasis. Although vertical lists serve some types of documents such as instructions and résumés, too many vertical lists disrupt the reading in journal articles and reports, in much the same way that traffic lights slow the driving through a city. Such disruptions make it difficult to read through longer documents. If the list is truly important, you might format it vertically. However, if you have more than one vertical list for every two or three pages of text, you should reconsider. Too many vertical lists will make your document appear like an outline.

Lists, particularly when they are long, are notorious for burying information. The following is a list of recommendations from Morton Thiokol to NASA on improvements needed for the solid rocket booster of the space shuttle. The list comes from a briefing that preceded the space shuttle *Challenger* disaster by over five months. Because the list was long, the emphasis given to the first recommendation was diluted [*Report*, 1986].

Recommendations

- The lack of a good secondary seal in the field joint is most critical and ways to reduce joint rotation should be incorporated as soon as possible to reduce criticality.
- The flow conditions in the joint areas during ignition and motor operation need to be established through cold flow modeling to eliminate O-ring erosion.
- QM-5 static test should be used to qualify a second source of the only flight certified joint filler material to protect the flight program schedule.
- VLS-1 should use the only flight certified joint filler material in all joints.
- Additional hot and cold subscale tests need to be conducted to improve analytical modeling of O-ring erosion problem.
- Analysis of existing data indicates that it is safe to continue flying existing design as long as all joints are leak checked with a 200 psig stabilization pressure, are free of contamination in seal areas and meet O-ring squeeze requirements.
- Efforts need to continue at an accelerated pace to eliminate SRM seal erosion.

In addition to the list being too long, Richard Feynman [1988] pointed out that there is a contradiction between the sixth item and the first.

How could the important details of this list be better emphasized? One improvement would have been to make a short list of the two or three most important recommendations followed by a list of the secondary recommendations on a separate page. Another improvement would have been to rework the language. The sentences are full of imprecision and needless complexity. For instance, in the first recommendation, what did the writer mean by the phrase "most critical"? There's no middle ground with the word "critical." Something is either critical or it isn't. Other problems with the language include wordiness, discontinuities, and needless passive voice.

References

Downing, R., "The MSAC Hotline," *Refrigeration Service and Contracting*, vol. 56, no. 9 (September 1988), p. 23.

Felder, Richard M., and James E. Stice, *National Effective Teaching Institute* (Toledo, Ohio: Annual Conference of the American Society for Engineering Education, 1992), p. 107.

Feynman, Richard P., "An Outsider's Inside View of the Challenger Inquiry," *Physics Today*, vol. 41 (February 1988), p. 29.

Foster, L., *Mountaineering Basics* (San Diego: Sierra Club and Avant Books, 1983).

Heath, D. H., and D. R. Williams, *Life at High Altitude* (London: Edward Arnold (Publishers) Limited, 1979).

Houghton, R. A., and G. M. Woodwell, "Global Climatic Change," *Scientific American*, vol. 260, no. 4 (April 1989), p. 42.

Lobenhoffer, H. P., R. A. Zink, and W. Brendel, "High Altitude Pulmonary Edema: Analysis of 166 Cases," *High Altitude Physiology and Medicine*, ed. by W. Brendel and R. Zink (New York: Springer-Verlag, 1982), pp. 219–231.

Poincaré, Jules Henri, *Science and Hypothesis*, trans. by G. B. Halstead (New York: Dover Publications, 1952), p. 141.

Radiation and Health Effects: A Report on the TMI-2 Accident and Related Health Studies, 2nd edition (Middletown, PA: GPU Nuclear Corporation, 1986), p. 1-1.

Reinhardt, W. G., and B. Groeneveld, "Phaseout of Chlorofluorocarbons Means Jobs, Gyrations for Builders," *Engineering News Report*, vol. 222 (June 8, 1989), pp. 8–10.

Reitz, K., "CFCs for Water Chillers," *Heating/Piping/Air Conditioning*, vol. 62, no. 4 (April 1990), pp. 57–61.

Report of the Presidential Commission on the Space Shuttle Challenger Accident, vol. 1 (Washington, DC: White House Press, 1986), p. 139.

Report of the Warren Commission on the Assassination of President Kennedy, New York Times Ed. (New York: McGraw-Hill, 1964), pp. 23–24.

Saathoff, Karl, "The Dangers of Breathing Compressed Air While Scuba Diving," *Undergraduate Engineering Review* (Austin, Texas: University of Texas, 1991).

Smith, M., "Avoiding High Altitude Illnesses," *Undergraduate Engineering Review* (Austin, Texas: University of Texas, 1993).

Ward, M., J. S. Milledge, and J. B. West, *High Altitude Medicine and Physiology* (Philadelphia : University of Pennsylvania Press, 1989).

Chapter 4

Language: Being Precise

When a writer conceives an idea he conceives it in a form of words. That form of words constitutes his style, and it is absolutely governed by the idea. The idea can only exist in words, and it can only exist in one form of words. You cannot say exactly the same thing in two different ways. Slightly alter the expression, and you slightly alter the idea.

—Arnold Bennett

In scientific writing, precision is the most important goal of language. If your writing does not communicate exactly what you did, then you have changed the work. One important aspect of precision is choosing the right word. Another important aspect of being precise is choosing the appropriate level of accuracy. Just as you wouldn't assign the wrong number of significant digits to a numeral in a calculation, so shouldn't you assign an inappropriate level of accuracy to details in your writing. The appropriate level of detail depends upon your work and your audience.

Choosing the Right Word

As a scientist or engineer, you wouldn't choose the word "weight" when you meant "mass." Technical terms such as "weight" and "mass" have specific meanings.

Many ordinary words have specific meanings as well. For instance, you shouldn't choose, as many writers do, the word "comprise" when you want the word "compose."

> Water is comprised of hydrogen and oxygen.

Because "comprise" literally means to embrace or to include, the above sentence is imprecise. A precise way to write this sentence is

> Water is composed of hydrogen and oxygen.

There are several word pairs that give difficulty not only to scientists and engineers, but to all writers. Listed below are some of these word pairs and the differences in their meanings.

affect/effect: *Affect* is a verb and means to influence (note that in psychology, *affect* has a special meaning as a noun). *Effect* is a noun and means a result (occasionally, people use *effect* as a verb meaning to bring about).

continual/continuous: *Continual* means repeatedly: "For two weeks, the sperm whales continually dived to great depths in search of food." *Continuous* means without interruption: "The spectrum of refracted light is continuous."

its/it's: *Its* is the possessive form of the pronoun *it* and is defined as "of it." *It's* is a contraction and is defined as "it is."

like/as: *Like* is a preposition and introduces a prepositional phrase. *As* is a conjunction and introduces a clause.

principal/principle: *Principal* can be either a noun or an adjective. As an adjective, *principal* means main or most important. *Principle* appears only as a noun and means a law ("Archimedes' Principle").

For a more detailed list see Appendix B.

Another precision problem, aside from that of selecting a word with an incorrect meaning, is grouping words such that they have no meaning:

> The problem centers around the drying and aging of the bloodstain.

The phrase "centers around" makes no physical sense. What the writer wants is either "centers on" or "revolves around."

> The problem revolves around the drying and aging of the bloodstain.

Many problems with word choice do not arise from confusion with other words or with lack of meaning. Rather, these problems arise because there is only one word that will do, and any substitute will cause imprecision. Despite what you may have learned, few words, if any, are exact synonyms. Some words have similar meanings, but are not interchangeable. Consider, for example, these two terms from the electrical breakdown of gases:

gas discharge: any one of the three steady states of the electrical breakdown process. These three steady states are the Townsend, glow, and arc.

spark: the transient irreversible event from one steady state of the electrical breakdown process to another (example: the transition from a glow discharge to an arc).

To use these two terms as synonyms is imprecise language. The first term is a steady state. The second term is a transition. However, in the first sentence of a recent journal article, a scientist tossed these terms around as if they were synonyms:

> The last decade has seen a rapid development of new techniques for studying the enormously complex phenomena associated with the development of sparks and other gas discharges.

Because a "spark" is not a "gas discharge," this sentence is imprecise. More important, because this sentence is the first sentence of the article, this imprecision undercuts the article's authority. Why did the scientist make this mistake? Given that the article was a review article and that the journal had solicited the scientist to write it, the scientist most likely didn't make the mistake because he didn't understand the vocabulary. He probably made the mistake because he was concentrating on the rhythm of the sentence rather than on the meaning. In other words, he wrote as if being fluid in the sentence was more important than being precise—something appropriate for poetry perhaps, but not for scientific writing.

Many scientists and engineers hold the misconception that using synonyms is a mark of a good writer. These scientists and engineers write with a pen in one hand and a thesaurus in the other. Well, synonym writing is not strong writing. Even when you find true synonyms, using them often confuses your readers:

> Mixed convection is a combination of natural and forced convection. Two dimensionless numbers in the correlations for mixed convection are the Grashoff number and the Reynolds number. The Grashoff number (for free convection) is a measure of the ratio of buoyant to viscous forces, and the Reynolds number (for forced convection) is a measure of the ratio of inertial to viscous forces.

Why did the engineer use "free convection" instead of "natural convection" in the third sentence? This synonym substitution added nothing to the discussion and only served to confuse readers unfamiliar with the vocabulary of heat transfer.

Another reason not to use synonyms is how inexact synonyms are. Consider how quickly these strings of synonyms, taken from a thesaurus, arrive at the antonyms of the first words:

perfect–pure–unvarnished–unfinished–rough–imperfect

valid–convincing–plausible–specious–unsound–invalid

classified–secret–mysterious–unidentified–unclassified

Most professional writers don't hesitate to repeat a word if that word is the right word. Moreover, most professional writers don't rely on a thesaurus. Most professional writers rely on dictionaries because dictionaries tell you the differences between words. Dictionaries help you find the right word.

I once had an English teacher who made us write five-sentence paragraphs about different subjects—subjects such as whales. In the paragraph, whales had to be the subject of each sentence, but you couldn't use the word "whales" more than once. You couldn't use pronouns, either. Instead, you had to come up with four synonyms for "whales." Which students did she praise? The students who gave up and decided that anything but whales was imprecise? No. She fell all over herself for the freckle-faced boy in the front row who wrote "king and queen mammals of the sea" and "black ellipsoids of blubber." With an evil eye for the rest of us, she said, "Now, he's a writer."

I have news for her—"black ellipsoids of blubber" is no kind of writing. Not scientific writing, not journalism, not fiction. You should test any writing advice you've received against the writing of great writers. Did Flannery O'Connor use a string of synonyms in her writing? Did Churchill? Did Einstein? No. They used the right word and only the right word. As Mark Twain said, "The difference between the right word and the almost right word is the difference between 'lightning' and 'lightning bug.'"

Besides the dictionary meaning (or denotation) of words, you should also worry about the connotation of words. A word's connotation is its associated meanings. Many words, such as "adequate," conjure an associated meaning that works against the dictionary meaning. In the dictionary, "adequate" means enough for what is required. However, would you fly on an airplane with an "adequate" safety record? Probably not, because "adequate" has a negative connotation. For many people, the connotation of "adequate" is the opposite of its denotation. When something is described as "adequate," many people think of that something as being insufficient. Another word with a strong negative connotation is "cheap." Although "cheap" has the same denotation as "inexpensive," the connotation differs. The connotation of "cheap" is that the item so described will not work.

Using a word with a negative connotation when a neutral or positive connotation is wanted is weak writing. So is the opposite, using a word with a positive connotation when a negative or neutral connotation is desired:

> The turbulence in the flow enhances the drag by more than 20 percent.

Because drag was an undesired quantity in this example, the engineer should have chosen a verb with a neutral or negative connotation ("enhance" has a positive connotation). Note that the denotation of "enhance" is also inappropriate here. To enhance something is to make an incremental change, but 20 percent is not incremental. A better word choice would have been either "increase" (neutral connotation) or "exacerbate" (negative connotation).

> The turbulence in the flow increases the drag by more than 20 percent.

Finally, in choosing the right word, go in fear of absolutes, especially in fear of the adverbs "always" and "never." Whenever you use these words, you invite your audience to look for exceptions, and if exceptions do exist, your audience will find them.

Choosing the Right Level of Detail

The previous section discussed precision from the viewpoint of choosing the right word. Precision also involves choosing the right level of accuracy for the details in your sentences. In scientific writing, you should strive not for the highest degree of accuracy, but for the appropriate level of accuracy.

How do you attain the appropriate level of detail for your language? One way is to achieve a balance between general statements and specific details. Strong writing requires both general and specific statements. General statements establish the direction of thought, and specific statements give evidence to support that direction. Writing, however, that relies solely on general statements is empty. Consider this stand-alone entry from a progress report to the Department of Energy:

> After recognizing the problems with the solar mirrors, we took subsequent corrective measures.

What were the problems with mirrors? What were the solutions? How many mirrors were damaged? This entry raises questions, but does not address them. Given that the field of solar mirrors cost over $40 million, this entry in the progress report did not satisfy the Department of Energy. A more precise entry to the report would have been as follows:

> Our last progress report (March 1985) discussed the damage to ten solar mirrors during a February thunderstorm. The question arose whether high winds or hailstones had cracked those mirrors. Now, after finding that high winds had caused the cracks, we have begun stowing all solar mirrors in a horizontal, as opposed to vertical, position during storms.

Another reason to use specific details is that general statements, by themselves, will not leave much of an impression on your audience. By themselves, generalities are soon forgotten.

> Our new process reduces emissions of nitrogen oxides from diesel engines and industrial furnaces.

Replacing this generality with a specific detail gives your audience something concrete to remember. Take a lesson from fiction writing. Good fiction writers rely on specific details to create scenes because good fiction writers know that specific details are what readers remember.

Our new process eliminates 99 percent of nitrogen oxide emissions from diesel engines and industrial furnaces.

Better yet, by grounding that detail (99 percent reduction), you insure that your audience understands its importance.

Our new process eliminates 99 percent of nitrogen oxide emissions from diesel engines and industrial furnaces. Previous processes have, at best, reduced nitrogen oxide emissions by only 70 percent.

Do not assume that general statements are inherently weak. In fact, sometimes general statements are desired. For instance, presenting specific details without general statements can be dangerous. Consider this example from an article on radon levels to a non-technical audience:

The average house in the area has a radon level of 0.4 picocuries per liter.

Without a general qualifier, the audience is left with the question, How dangerous is 0.4 picocuries? That kind of writing is not only weak, but irresponsible. Revision gives

The average house in the area has a radon level of 0.4 picocuries per liter, which is considered low by the EPA [Lafavore, 1987]. Levels between 20 and 200 picocuries per liter are considered high, and levels above 200 picocuries per liter are considered dangerous. For reference, the average radon level in outdoor air is about 0.2 picocuries per liter.

Sometimes specific details confuse because they give too much information. In other words, the specific details raise side issues for readers that the writer does not intend.

The number of particular hydrocarbon combinations in our study is enormous. For example, the number of possible $C_{20}H_{42}$ is 366,319 and the number of $C_{40}H_{82}$ is 62,491,178,805,831.

What was the purpose of including these two numbers? The chemist wanted to show the extent of his calculations. However, were all those digits necessary? Also, because the second number was so much larger than the first number, was the first number necessary? A more precise paragraph would have been as follows:

> The number of hydrocarbon combinations in our study is enormous. For example, the number of possible $C_{40}H_{82}$ is over 60 trillion.

This revision achieves the desired result—showing the extent of the calculations—without making the audience wade through undesired numerals.

Consider another example, this one from a progress report about a solar power plant:

> Operations at the plant stopped momentarily because the thermal storage charging system desuperheater attemperator valve was replaced.

The name "thermal storage charging system desuperheater attemperator valve" is a problem. For one thing, when written as a single noun phrase, it is difficult to read. For another thing, the report's readers (plant managers) did not know exactly what this particular valve was. All the readers knew about this valve was that it was in the thermal storage system. For that reason, a more appropriate level of accuracy would have been as follows:

> Operations at the plant stopped for 1.5 hours so that a valve in the thermal storage system could be replaced.

In this revision, the location of the valve was made less specific, while the time that the plant was down was made more specific. This report's readers cared more about how long the plant was down than about which particular valve was leaking.

Packing sentences with too many details also makes for tiresome reading:

> A 1-mm diameter, 656-nm beam with uniform intensity across the beam was produced by using a wavelength/polarizer combination to split off part of the 532-nm output from the Nd:YAG laser to pump a second dye laser (Laser-

> Ray LRL-2, also operated with the DCM dye) with a side-pumped configuration for the final amplifier, and selecting the central portion of the collimated beam with an aperture.

There are just too many details in this one sentence. Are all these details necessary? Couldn't the physicist have spread these details over several sentences? Better still, couldn't the physicist have placed the secondary details (such as beam wavelength and manufacturer's name) in an illustration? Scientists and engineers sometimes worry so much about telling readers everything that they end up not informing readers of anything.

Being precise doesn't mean compiling details; being precise means selecting details. You should choose details that inform:

> The fuel pellets used in inertial confinement fusion are tiny, the size of BBs, but they are potentially the most powerful devices mankind has ever known. If we can compress the fuel in the pellets to a plasma, the fuel's deuterium and tritium atoms can overcome their mutual electrical repulsion and fuse into helium atoms, giving off energy ($E = mc^2$). The power needed to ignite fusion in the pellets is 100 trillion watts; however, the power released from the fusion is one hundred times that much.

This paragraph informs; it informs because the scientist selected the most important details about the pellets. Because scientific writing is compressed, you have room for only the most important details. Make them count.

References

Bennett, A., *Literary Taste and How to Form It* (London: George H. Doran Publishers, 1909).

LaFavore, M., *Radon, the Invisible Threat* (Pennsylvania: Rodale Press, 1987).

Chapter 5

Language: Being Clear

When you are out to describe the truth, leave elegance to the tailor.

—Albert Einstein

While precision in scientific writing means saying what you mean, clarity means avoiding things that you don't mean. Too often in scientific writing, a cloudy sentence disrupts the continuity and authority of an entire section. In scientific writing, each sentence builds on the ones around it. If one sentence is weak, your language falters and your readers stumble. What makes writing unclear? In scientific writing, there are two principal sources: ambiguity and needless complexity.

Avoiding Needless Complexity

If there were only one piece of stylistic advice I could whisper into the ear of every scientist and engineer, it would be to "keep things simple." Undoubtedly, the worst language problem in scientific writing is needless complexity. This needless complexity arises in words, phrases, and sentences.

Needlessly Complex Words. Many words used in scientific writing add no precision or clarity to the writing—only complexity. Table 8-1 presents examples of needlessly complex *nouns, verbs, adjectives,* and *adverbs*. Most of these words have distinguishing traits. For example, many needlessly complex verbs end in *-ize*. Although there are words such as "minimize" and "maximize" that have this ending and still are clear, you should recognize words with such distinguishing traits and challenge them in your writing. Are the words precise? Are they clear? If so, then use them. If not, find simpler substitutes.

Table 8-1
Examples of Needlessly Complex Words

Category	Example	Possible Substitute
nouns	familiarization	familiarity
	has the functionability	can function
	has the operationability	can operate
	utilization	use
verbs	facilitate	cause
	finalize	end
	prioritorize	assess
	utilize	use
adjectives	aforementioned	mentioned
	discretized	discrete
	individualized	individual
	personalized	personal
adverbs	firstly, secondly, thirdly	first, second, third
	heretofore	previous
	hitherto	until now
	therewith	with

You might think that these substitutions don't make that much difference, and individually each substitution does not. Collectively, though, these substitutions have a profound effect. Consider an example in which the writer consistently chooses the needlessly complex word:

> The objective of this study is to develop an effective commercialization strategy for solar energy systems by analyzing the factors that are impeding early commercial projects and by prioritizing the potential government and industry actions that can facilitate the viability of the projects.

This sentence is inaccessible to readers. Are the words "commercialization," "prioritizing," "facilitate," and "viability" necessary? Revision with attention to precision and clarity gives

> This study will consider why current solar energy systems have not yet reached the commercial stage and will evaluate the steps that industry and government can take to make these systems commercial.

Opting for the simpler word choices does not make your writing simplistic—you've got enough inherent complexity in your ideas to surpass that label. Rather, opting for the simpler word choices makes your ideas clear to your readers.

Needlessly Complex Phrases. Just as words in scientific writing are often needlessly complex, so too are phrases. One source of needless complexity in phrases comes from stringing modifiers in front of nouns. Nouns, particularly subject nouns, are stepping stones in sentences, and stringing modifiers (adjectives and other nouns acting like adjectives) in front of them confuses readers. Readers don't know where to step.

> Solar One is a 10 megawatt solar thermal electric central receiver Barstow power pilot plant.

Imagine the engineer who wrote this sentence actually saying it out loud. He would lose his breath before finishing. If all these modifiers are important, then the engineer should either place them in phrases and clauses around the principal noun "plant" or work them into two sentences:

> Solar One is a solar-powered pilot plant located near Barstow,

> California. Solar One produces 10 megawatts of electric power by capturing solar energy in a central receiver design.

Stringing modifiers dilutes the meanings of the modifiers. The modifiers become lost in the string. Moreover, long strings bring imprecision into sentences:

> The decision will be based on economical fluid replenishment cost performance.

What exactly does the phrase "economical fluid replenishment cost performance" mean? Will the decision be based on the performance of the fluid or on the cost of replacing the fluid or on something else? Pursuing our goals of precision and simplicity, we revise this sentence to

> We will base the decision on the cost of replacing the thermal oil.

Needlessly Complex Sentences. In many scientific documents the sentences are needlessly complex. For instance, the sentence lengths in many scientific documents average over thirty words, as compared to an average length of less than twenty in most newspapers. More important than length is that many sentence structures in scientific writing are convoluted:

> The object of the work was to confirm the nature of electrical breakdown of nitrogen in uniform fields at relatively high pressures and interelectrode gaps that approach those obtained in engineering practice, prior to the determination of the processes that set the criterion for breakdown in the above-mentioned gases and mixtures in uniform and non-uniform fields of engineering significance.

This sentence is complex. It is needlessly complex. It has sixty-one words, two of which are hyphenated. What's more important, the sentence structure is convoluted. After the word "gaps," the sentence wanders aimlessly from one prepositional phrase to another. Prepositional phrases are important; they incorporate details of time, manner, and place. This sentence, however, has eleven prepositional phrases. Count them—eleven. Prepositional phrases act in the same way as boxcars on a train. They provide no momentum, only friction. Something else that hurts this sentence is that the reader has no clue that this sentence will be long. The sentence could logically end after the thirteenth word ("breakdown"), but the sentence doesn't. It just wanders.

Would the sentence be strengthened by breaking it up into shorter sentences? Yes, but don't take shorter sentences as a cure for all unclear writing. Although you could strengthen many scientific documents by making all the sentences short, using only short sentences will not produce strong writing, especially in a longer document. You also need long and medium-length sentences to provide variety and emphasis. Moreover, it is not long sentences that confuse readers; it is complex sentences that confuse readers. Consider a revision for the first half of the sentence:

> At relatively high pressures (760 torr) and typical electrode gap distances (1 mm), the electrical breakdown of nitrogen was studied in uniform fields.

This revised sentence has twenty-three words, which is a relatively long sentence. Nevertheless, the sentence is successful. Instead of eleven prepositional phrases, the sentence has only three. Notice that the revision anchors the sentence's general qualifiers with specific examples.

How do you know when a sentence is too complex? One signal is when the sentence contains too many ideas.

> To separate the hot and cold oil, one tank was used that took advantage of the thermocline principle, which uses the rock and sand bed and the variation of oil density with temperature (8% decrease in density over the range of operating temperatures) to overcome natural convection between the hot and cold regions.

Does the sentence inform? No. This sentence is indigestible. The writer has presented the reader with several ideas, but in a single sentence these ideas cluster into lumps as oatmeal does when it hasn't been stirred.

Before fixing this sentence, you should decide which details in the sentence are important. Does the reader really need to know how the hot and cold oil are separated or how the density varies in the tank or what the thermocline principle is? If these details are not important, then you should delete them. However, if the reader needs to know these details to understand the work, then you should present them in digestible portions.

> The question was how to separate the hot and cold oil in the rock and sand bed. Rather than have one tank hold hot oil and another tank hold cold oil, we used a single tank for both. This design took advantage of the variation of oil density with temperature. In our storage system, oil decreases in density by 8% over the range of operating temperatures. This variation in density allows the hot oil to float over the cold oil in the same tank. Natural convection is impeded by the position of the hot oil over the cold and by the rock and sand bed. Thus, the heat transfer between hot and cold regions is small because it occurs largely by conduction. The concept of storing heat in a single vessel with hot floating over cold is known as "thermocline storage."

In evaluating whether a sentence is too complex, you should make sure that the sentence contains no more than one main idea and that the sentence doesn't wander. Perhaps a better test for deciding whether a sentence is too complex is if you notice that the sentence is long. In a well-written long sentence, you don't notice the length; rather, you just take in the information and move on.

Besides having sentences that are too long, much scientific writing suffers because the structures of the sentences are antiquated. Writing styles change from generation to generation. For instance, although Theodore Dreiser's writing style was accepted at the beginning of the twentieth century, his style is outmoded today. As a scientist or engineer, you should choose your writing models carefully. Great scientists and engineers are not necessarily great writers. Consider a discussion written by the great physicist Niels Bohr [1924]:

> ***The Correspondence Principle.*** So far as the principles of the quantum theory are concerned, the point which has been emphasized hitherto is the radical departure from our usual conceptions of mechanical and electrodynamical phenomena. As I have attempted to show in recent years, it appears possible, however, to adopt a point of view which suggests that the quantum theory may, nevertheless, be regarded as a rational generalization of ordinary conceptions. As may be seen from the postulates of the quantum theory, and particularly the frequency relation, a direct connection between the spectra and the motion of the kind required by the classical dynamics is excluded, but at the same time the form of these postulates leads us to another relation of a remarkable nature.

This writing is needlessly complex. There are many bloated phrases such as "which has been emphasized hitherto." The sentences are also long. They average almost forty words. More important, the sentence structures are convoluted. "Ah," you say, "but this writing is elegant and beautiful." Is it? The ideas are beautiful, but the writing is murky, almost inaccessible. In scientific writing, beauty lies in clarity and simplicity:

> ***The Correspondence Principle.*** Many people have stated that the quantum theory is a radical departure from classical mechanics and electrodynamics. Nevertheless, the quantum theory may be regarded as nothing more than a rational extension of classical concepts. Although there is no direct connection between quantum theory postulates and classical dynamics, the form of the quantum theory's postulates, particularly the frequency relation, leads us to another kind of connection, one that is remarkable.

When writing, you should imagine yourself sitting across from your most important reader. Write your paper as if you were talking to that reader. This exercise doesn't mean that your writing should be informal.

Rather, it means that you should rid your writing of needless formality. The purpose of scientific writing is to inform, not to impress. Therefore, don't write

> In that the "Big Bang," currently the most credible theory about how the universe was created, explains only the creation of hydrogen and helium, we are left to theorize as to how all the other elements came into being. Having studied the nuclear reactions that constitute the life and death cycles of stars, many scientists believe therein lies the key.

This kind of writing bewilders. The "in that" clause that begins the first sentence and the long *participial phrase* that begins the second sentence are outmoded sentence structures. If you need a writer to emulate, choose Richard Feynman. He probably would have written the paragraph along these lines:

> The "Big Bang" is the most credible theory for the creation of the universe. Nevertheless, the "Big Bang" explains the creation of only helium and hydrogen. What about the other elements? Many scientists believe that they arose from nuclear reactions that occur in the life and death cycles of stars.

Avoiding Ambiguity

Besides avoiding needless complexity, you also should avoid ambiguity in your writing. Ambiguity is created by the use of a word, phrase, or sentence that can be interpreted in more than one way. Many ambiguities in scientific writing are difficult to classify.

> The solar collector worked well under passing clouds.

Does the solar collector work at a height that is well below the passing clouds, or under passing clouds does the solar collector work well? Your English teacher would probably say that this ambiguity is a careless writing mistake. Perhaps it is, but it's not so much a careless writing mistake as a careless editing mistake. Everyone makes mistakes on early drafts. What makes writing strong is not some dose of magic conjured on the first draft. Rather, strong writing arises from hard work on revisions.

Consider another editing oversight, this one from a section presenting a design of a system to measure the radiation striking a solar energy receiver:

> ***Radiometer System.*** Two general requirements to be met are (1) to survive and accurately measure the radiation incident on the receiver and (2) to present the data in a form that can be used to verify computer code predictions.

The engineer did not smooth the "trail" here. First, the transition into the section is abrupt. The writer should include the words "radiometer system" in the first sentence of the section. Second, the sentence has gaps. Who or what must survive? Finally, the engineer says that there are two requirements, yet lists three. Attention to precision and clarity gives this revision:

> ***Radiometer System.*** The radiometer system for the solar receiver must meet three requirements: (1) it must accurately measure to within 5 percent the solar radiation on the receiver; (2) its electronics must survive in solar radiation as intense as 300 kilowatts per square meter; and (3) its output must be able to verify computer codes.

Note that there are many successful ways to rewrite the original sentence. The point is that the sentence needs rewriting.

Some scientists and engineers lament about how tired they are after writing a first draft. "I can't look at it anymore," they say. "It's too painful." Perhaps it is painful, especially if it was drafted with the same imprecision as in the example above. However, the writing represents your work, and if you spent time on your work, then you should also spend time on documenting that work. You do not attain clarity on first drafts when you have a blank computer screen staring at you. You attain clarity on final drafts when you have your work—your ideas and calculations—on paper in front of you. Although the ambiguities in the examples so far are difficult to classify, many ambiguities arise from four specific sources: word choice, syntax, pronouns, and punctuation.

Ambiguities in Word Choice. Many words in English have multiple meanings. For instance, the word "as" has several denotations. Although it is outdated to use "as" to mean "because," many scientists and engineers still try to use the word "as" in this older sense. The result is often an ambiguity:

> T cells, rather than B cells, appeared as the lymphocytes migrated to the thymus gland.

Although the author intended the word "as" to mean "because," many readers would interpret "as" to mean "while." Rather than use the word "as" in this sentence, the writer should use the word "because."

> T cells, rather than B cells, appeared because the lymphocytes migrated to the thymus gland.

Ambiguities in Syntax. Some ambiguities arise because of errors in syntax. Syntax refers to the order and structure of words or phrases in a sentence. When scientists and engineers are not careful about the ordering of words and phrases, ambiguities can occur. For example, consider how the placement of "only" can change the meaning of a sentence:

> *Only* I tested the bell jar for leaks yesterday.
> I *only* tested the bell jar for leaks yesterday.
> I tested *only* the bell jar for leaks yesterday.
> I tested the bell jar *only* for leaks yesterday.
> I tested the bell jar for leaks *only* yesterday.

This example presents five different syntactical arrangements of this sentence, and each sentence has a different meaning. Be careful about the placement of words, particularly the placement of words such as "only" and "well" that are adverbs in some situations and adjectives in others.

The improper placement of phrases, particularly introductory phrases, can also cause ambiguities.

> In low water temperatures and high toxicity levels of oil, we tested how well the microorganisms survived.

I hope that everyone conducting the tests survived as well. Revision with attention to clarity gives

> We tested how well the microorganisms survived in low water temperatures and high toxicity levels of oil.

Ambiguities in Pronouns. Many ambiguities arise because of mistakes with pronouns. According to Fowler's *A Dictionary of Modern English Usage* [1965], an important principle for using pronouns is that there should not be "even a momentary doubt" as to what the pronoun refers. Many scientists and engineers, unfortunately, ignore this principle. They abuse pronouns, particularly the pronouns "it" and "this."

> Because the receiver presented the radiometer with a high-flux environment, it was mounted in a silver-plated stainless steel container.

What is mounted in the container? The receiver? The radiometer? The environment? The noun "environment" is the nearest possible reference to the pronoun "it," but that reference makes no sense. The noun "receiver" receives the most emphasis in the sentence, but three pages later in this report the readers learn that the receiver stands over 45 feet tall, making it much too large to fit into the container. In other words, the engineer had intended the "it" to refer to the noun "radiometer," although this noun was neither the noun nearest to the pronoun nor the noun that received the most emphasis in the sentence. Given the number of possible references in this example, the engineer should have just repeated the noun "radiometer," rather than use the pronoun "it."

The way that many scientists and engineers treat the pronoun "it" is unsettling, but the way that many scientists and engineers treat the word "this" is criminal:

> There are no peaks in the olefinic region. Therefore, no significant concentration of olefinic hydrocarbons exists in fresh oil. This places an upper limit on the concentration of olefins, no more than 0.01 percent.

What does the chemist want the "this" to refer to? To the last noun of the previous sentence—oil? To the subject of the last sentence—concentration? To the idea of the previous sentence—that there is no significant concentration of olefinic hydrocarbons? Actually, the "this" in this example refers to none of these. The chemist intended the "this" to refer to the lack of peaks in the olefinic region.

Unlike "it," which is a pronoun and nothing else, the word "this" is an adjective, which some writers informally use as a pronoun. "This" is a directive, a pointer. Using "this" in a sentence as an anchor (as a true pronoun) often weakens the writing. What does the "this" point to? The last noun used? The subject of the last sentence? The idea of the last sentence? Or something else? Instead of using "this" as an anchor, clarify your writing by letting "this" do what it does best—point.

> The chromatogram has no peaks in the olefinic region. Therefore, no significant concentration of olefinic hydrocarbons exists in fresh oil. This chromatogram finding places an upper limit on the olefin concentration, no more than 0.01 percent.

Ambiguities in Punctuation. A fourth common source of ambiguities is punctuation. Punctuation marks act as road signs in the writing. The marks tell readers when to stop and when to slow down. The punctuation mark that causes the most ambiguities is the comma. Commas act as yield signs that tell readers when to slow down so that they will see the sentence in a certain way.

There are many rules for commas. Some rules are mandatory, others not. When a comma is needed to keep a sentence from being misread, then it is mandatory. Consider an example in which someone did not follow this rule:

> After cooling the exhaust gases continue to expand until the density reaches that of freestream.

This sentence requires a comma after "cooling"; otherwise, readers don't know where the initial phrase stops and the main sentence begins. Consider another example:

> When feeding a shark often mistakes undesirable food items for something it really desires.

Here, the scientist needs a comma after "feeding." If you are in doubt about whether to place a comma after an introductory phrase, then go ahead and place it. There's nothing wrong with the comma being there, even when it isn't necessary.

Besides the absence of commas after introductory phrases, another mistake with commas resulting in ambiguity occurs in a series of three or more items. The "old" punctuation rule for commas in a series is that in a series of three or more items, you should use commas to separate each term. For example,

> The three elements were hydrogen, oxygen, and nitrogen.

Because of influences from literature and journalism, this rule has changed over the years to the point where the comma after "oxygen" is optional as long as there is no ambiguity. The problem is that in scientific writing there often are ambiguities. Consider this sentence from a recent journal article:

> In our study, we examined neat methanol, neat ethanol, methanol and 10 percent water and ethanol and 10 percent water.

The ten-thousand dollar question is how many fuels did the chemist examine. Three? Four? Five? A good lawyer could make a case for at least three different answers. In fact, lawyers often make cases showing ambiguities that scientists and engineers have written. As long as an ambiguity exists, the scientist or engineer loses. To rid the above example of the ambiguity, the writer would benefit from using a colon and a return to the old-fashioned way of punctuating items in a series:

> In our study, we examined four fuels: neat methanol, neat ethanol, methanol with 10 percent water, and ethanol with 10 percent water.

As stated, ambiguities are difficult to find when you're familiar with the writing. One thing that will help you is to get some distance from your draft. Not only will time accomplish that, but so will a different setting for reading the document. In addition to reading the document in a different room other than the one in which you drafted it, you can find a different setting by printing out the document in a different format (for instance, double space rather than single space). The different format will give the document a new look. Better still, you should try to find two or three conscientious readers for the document. If there's an ambiguity, chances are that one of them will catch it.

References

Bohr, Niels, *The Theory of Spectra and Atomic Constitution* (Cambridge: Cambridge University Press, 1924), p. 81.

Einstein, Albert, (1912) qtd. in *Albert Einstein: eine dokumentarische Biographie*, by Carl Seelig (Vienna: Europa-Verlag, 1954), p. 168. Original text: *"Die Schönheit, meine Herren, wollen wir den Schustern und Schneidern überlassen. Unser Forschungsziel muß die Wahrheit bleiben."*

Fowler, H. W., *A Dictionary of Modern English Usage*, 2nd ed. (Oxford: Oxford University Press, 1965), p. 481.

Chapter 6

Language: Being Forthright

Short words are best, and old words, when short, are best of all.
—Winston Churchill

The purpose of scientific writing is to inform. Therefore, the attitude of the writer should be forthright; it should be sincere and straightforward. To be forthright in your writing, you have to control your tone. Tone is whatever in your language indicates the attitude that you, the writer, have towards your subject. Also, to be forthright you should select strong nouns and verbs. Nouns and verbs are the most important words in sentences. When your nouns and verbs are weak, your writing becomes lethargic.

Controlling Tone

Tone is whatever in the writing shows your attitude toward the subject. Often, scientists and engineers lose control of tone by avoiding simple, straightforward wording. When some people sit down to write, they change their entire personalities. For instance, in correspondence, instead of using normal English, many scientists and engineers use stilted phrases such as "per your request" and "enclosed please find." Because these phrases are not straightforward, they inject an undesired attitude—often of pretentiousness or arrogance—into the writing.

Avoiding Pretentious Words. Many times, in order to be precise, you use unusual words. Words such as "polystyrene" and "glycosuria" have no simple substitutes, but many words used in scientific writing are needlessly unfamiliar. Words such as "facilitate" and "implement" are pretentious. As stated in the previous chapter, these words offer no precision or clarity to the writing. Worse yet, they smack of a pseudo-intellectuality that poses a barrier between writer and reader. Some of these words ("interface," for instance) are precise in certain contexts, yet imprecise and pretentious in others. An interface is the interstitial boundary between two molecular planes, and when used in this way, "interface" is precise. However, when scientists and engineers use "interface" instead of the verb "to meet," an awkward image occurs.

> In the development of the Five-Year Plan, Division 8462 will interface with Division 8573.

Two groups of people actually "interfacing" suggests interactions that are, well, unprofessional.

Below is a short list of common pretentious words found in scientific writing:

approximately: appropriate when used to modify a measurement's accuracy to within a fraction, but pretentious when applied to a rougher estimation such as "approximately twelve people." Does the writer mean 11.75 or 12.25 people? In such cases, use the simple word "about."

component: often can be replaced by *part*.

facilitate: a bureaucrat's word. Opt for the simpler wording *cause* or *bring about*.

implement: a farm tool. Many writers, however, have followed the bureaucratic winds and now use *implement* as a verb. I suggest using *put into effect* or *carry out*. These verb phrases are old and simple. They're the verb phrases Winston Churchill would have used.

manufacturability: a pretentious way to say "can manufacture." You should challenge *-ability* words; their presence often indicates that the sentence should be rewritten with a stronger verb preceded by the word *can*.

utilize: a pretentious way to say the verb *use*. You should challenge *-ize* verbs. Many times, there is a simpler substitute.

utilization: a pretentious way to say the noun *use*. You should challenge *-ization* nouns.

Many of these words are long, most of these words are recent creations, and all of these words have infested scientific writing. What these kinds of words do is to hide the meanings of sentences:

> The elevated temperatures of the liquid propellant fuel facilitated an unscheduled irrevocable disassembly.

What actually occurred in this seemingly impregnable sentence was straightforward.

> The higher temperatures caused the liquid propellant to explode.

Avoiding Arrogant Phrases. Besides using pretentious words, many scientists and engineers lose control of tone by choosing arrogant phrases.

> As is well known, the use of gaseous insulation is becoming increasingly more widespread, with gases such as air and sulphur hexafluoride featuring prominently. There has been some discussion of using gas mixtures like nitrogen and sulphur hexafluoride, and of course nitrogen is the major constituent of air.

Besides problems with precision and clarity, these two sentences have problems with tone. The phrase "as is well known" is not forthright. If the readers know (and know well) the particular detail about gaseous insulation, then why does the writer mention it? But if the readers do not know the detail, as is probably the case, then the writer has assumed a superior position over the reader. The phrase "of course" in the second sentence strikes the same chord.

Another arrogant phrase often found in scientific writing is "clearly demonstrate."

> The results clearly demonstrate the ability of Raman spectroscopy to provide unambiguous chemical compound identifications from oxides as they grow on a metal surface.

When someone uses "clearly demonstrate" to describe results or illustrations, more often than not those results or illustrations don't "clearly demonstrate" anything at all. Therefore, readers are left wondering if the writer is trying to hide something. The word "unambiguous" in the same example is also arrogant; it defies the reader to question the figure. A forthright revision would be

> The results show that Raman spectroscopy can identify chemical compounds from oxides that are growing on a metal surface.

The revision is simple and straightforward. Scientific work should stand on its own merit without cajoling from the writer.

Perhaps the most arrogant expression in scientific writing is the phrase "it is obvious." If the detail is obvious, then you shouldn't include it, and if the detail is not obvious, as is usually the case, calling the detail obvious will serve only to annoy your readers. Many insecure scientists and engineers hide behind these arrogant phrases. These scientists and engineers think—subconsciously, perhaps—that the arrogant phrases will ward off challenges to their work. No such luck. If anything, these phrases invite rebuttal.

Avoiding Silliness. Rather than using pretentious words or arrogant phrases, some scientists and engineers incorporate a silliness into their writing. The most common way this silliness arises is through clichés. Clichés are figurative expressions that have become too familiar. These expressions were once fresh in writing, but through overuse, have taken on undesirable connotations.

> When you come up to speed, I will touch base with you and we'll knock heads together and figure out a solution!

"Come up to speed," "touch base with you," "knock heads together"—these phrases are overused to the point of sounding "cute." The exclamation point also falls into this category. So many *Dick and Jane* readers have used the exclamation point for emphasis that it doesn't emphasize anymore. Rather, the exclamation point conjures images of *Dick and Jane* readers.

Clichés are often found in scientific correspondence. Consider the following paragraph from a job application letter (note that the ellipsis was part of the original paragraph):

> I am in search of a meaningful growth position, and your company is a gateway to the highest professional levels and beyond...

The clichés in this sentence are imprecise and unclear. The phrase "meaningful growth position" sounds painful, the word "gateway" suggests that the writer wants a job in a port city such as St. Louis, and the phrase "and beyond" followed by an ellipsis ("...") conjures an opening to a space adventure movie. While the ellipsis is appropriate when you are indicating something missing from a quotation, it is not appropriate as a pointed finger to the future (too many Harlequin romances have ended this way).

How do you know whether a descriptive phrase is a cliché? If you've heard the descriptive phrase several times before ("sticks out like a sore thumb") or if it strikes a high-pitched chord that makes you cringe ("greatest thing since sliced bread"), then it's probably a cliché.

Choosing Strong Nouns and Verbs

Nouns and verbs are the most important words in sentences. Nouns provide stepping stones in sentences, and verbs provide momentum. When your nouns and verbs are weak, your writing is lethargic. When your nouns and verbs are strong, your writing is crisp.

Using Strong Nouns. When a noun is strong, it provides one of the five senses to the reader—most often, a visual image. When a noun is weak, it does not provide any of the five senses.

> The existing nature of Mount St. Helens' volcanic ash spewage was handled through the applied use of computer modeling capabilities.

The nouns in this example—"nature," "spewage," "use," and "capabilities"—are weak. Why? "Nature," "use," and "capabilities" provide no helpful images or any other sensations to the readers. "Nature" may provide an image, such as a brook or a deer, but this image doesn't clarify the meaning of the sentence. The fourth noun, "spewage," isn't really a noun, but a verb ("to spew") cloaked as a noun. Taking a verb and twisting it into a noun is similar to taking a baseball pitcher and sending him up to bat in a critical situation. Granted, there are some pitchers (Bob Gibson, for example) who hit well, but most pitchers are easy outs. Just as a baseball manager uses hitters to hit and pitchers to pitch, so should a writer use nouns as nouns and verbs as verbs. Because the four nouns in the example are weak, the sentence suffers. Revision gives

> With Cray computers, we modeled how much ash spewed from Mount St. Helens.

This revision is much crisper, more direct. Notice that the nouns in this revision—"computers," "ash," and "Mount St. Helens"—provide images for the readers.

Nouns that provide one of the five senses are called *concrete nouns*, and nouns that do not provide one of the five senses are called *abstract nouns*. In general, reducing the number of abstract nouns will strengthen your writing. Here is a short list of abstract nouns that often creep into scientific writing:

ability	environment
approach	factor
capability	nature
concept	parameter

"Environment" is particularly bothersome. When used to discuss the world's ecosystem, the word is appropriate. However, some scientists and engineers tag the word onto every other noun. These scientists and engineers don't talk about the "laboratory" or "computer" anymore, rather the "laboratory environment" or the "computer environment." Recently, I heard someone discussing the "farm environment." What's wrong with the word "farm"? Adding the word "environment" in such a situation is superfluous and only serves to cloud the sentence's meaning.

Notice that an abstract noun such as "methods" or "criteria" can serve a document so long as the writer anchors the abstract noun with a concrete definition or an example:

> This study considered three methods for detecting plastic explosives in airline baggage: ion mobility spectrometry, x-ray backscatter, and thermal neutron activation. To evaluate these three methods, we used four criteria: cost of the method, effectiveness of the method at identifying a plastic explosive, speed of the method at processing baggage, and ease of use for the method.

Because the writer grounded the words "methods" and "criteria" with concrete examples, the writer made these abstract nouns concrete for the remainder of the document.

Using Strong Verbs. Verbs provide momentum in sentences. When your verbs are strong, your writing moves. Many scientists and engineers sap strength from their verbs by burying them as nouns in weak verb phrases:

Weak Verb Phrase	Strong Verb
made the arrangement for	arranged
made the decision	decided
made the measurement of	measured
performed the development of	developed

Weak verb phrases make your sentences lethargic:

> The human immune system is responsible not only for the identification of foreign molecules, but also for actions leading to their immobilization, neutralization, and destruction.

In forthright writing, the words that contain the natural action of the sentence serve as the verbs:

> The human immune system not only identifies foreign molecules, but also immobilizes, neutralizes, and destroys those molecules.

While the original sentence just sits, the revised sentence moves. It moves because the verbs are active. The active verbs "identifies," "immobilizes," "neutralizes," and "destroys" have replaced the passive verb "is."

Sometimes, scientists and engineers bury strong verbs in other constructions:

Buried Verb	Strong Verb
is beginning	begins
is following	follows
is shadowing	shadows
is used to detect	detects

These constructions slow the reading:

> The distance between the electrodes is dependent on the pressure environment and the result of metal corrosion.

The verb of this sentence is "is." The verb "is," which is a form of the verb "to be," is important in scientific writing. It acts as an equal sign, which you need for sentences that define or equate:

> A *positron* is a positively charged electron produced in the beta decay of neutron-deficient nuclides.

However, if something other than a definition or an equality occurs in the sentence, then you should look to another verb besides "to be." In the previous example, this other verb is buried in the adjective "dependent." Revision gives

> The distance between the electrodes depends on the gas pressure and the electrodes' corrosion.

Note how changing the verb from "is" to "depends" exposes the abstract nouns "environment" and "result."

Many scientists and engineers hold the misconception that scientific documents should be written in the *passive voice*. Not true. Because the purpose of scientific writing is to communicate (inform or persuade) as efficiently as possible, and because the most efficient way to communicate is through straightforward writing, you should use the most straightforward verbs available. Needlessly passive verbs slow your writing; they reduce your writing's efficiency.

> The feedthrough was composed of a sapphire optical fiber, which was pressed against the pyrotechnic that was used to confine the charge.

Eliminating the passive voice from this sentence strengthens the writing:

> The feedthrough contained a sapphire optical fiber, which pressed against the pyrotechnic that contained the charge.

Is all passive voice wrong? No. Although the *active voice* ("The oscilloscope displayed the voltage") is stronger than the passive voice ("The voltage was displayed on the oscilloscope"), there are occasions when the passive voice is more natural. For instance,

> On the second day of our wildebeest study, one of the calves wandered just a few yards from the herd and was attacked by wild dogs.

In this example, there is nothing wrong with the passive verb "was attacked" because the passive voice allows the emphasis to remain on the wildebeest calf, which is the focus of the paragraph.

The key to choosing between an active and passive verb is to ask which form is more natural. Most often, an active verb is more natural. Unfortunately, many scientists and engineers opt solely for passive verbs. Unnatural passive voice is lifeless. If energy has occurred in the work, unnatural passive voice saps that energy and leaves dead words on the page.

> A new process for eliminating nitrogen oxides from diesel exhaust engines is presented. Flow tube experiments to test this process are discussed. A chemical reaction scheme to account for this process is proposed.

This kind of writing could cure insomnia. The work isn't boring—in fact, it's engaging—but the writing is boring. The writing is boring because all the verbs are passive, and most passive voice is unnatural to our ears. Imagine listening to a baseball game in which the radio announcers speak in an unnatural passive voice:

> The ball is pitched by Gibson. The bat is swung by Clemente. The ball is hit by Clemente. The outfield is covered by Brock. The fence is climbed by Brock.

Boring, boring, boring. Radio announcers wouldn't call a game that way—if they did, they wouldn't have many listeners. A good baseball announcer (Harry Carey, for example) would call the same play in a natural active voice, so that the listener can experience the energy of the event:

> Clemente digs in. Gibson comes set, delivers. Clemente swings. He hits a deep drive to left. Brock goes back, way back. He climbs the wall. Kiss that baby good-bye.

Just as we strengthened the radio broadcast by using the active voice, so too can we strengthen the article summary.

> This paper presents a new process for eliminating nitrogen oxides from the exhaust of diesel engines. To test this process, we performed experiments in flow tubes. To explain this process, we developed a scheme of chemical reactions.

How do you take needlessly passive writing and make it active? This question has no simple answer. One way to circumvent the passive voice is to allow objects to do the things that they're endowed to do. Because many readers feel uncomfortable with inanimate objects acting in sentences, you have to be careful. Suppose in the first draft of a section about using an oscilloscope, you write

> The voltage was given on the oscilloscope.

and you want to convert that sentence to the active voice. Which active verb you choose determines the reader's comfort level with the inanimate object's action.

The oscilloscope displayed the voltage.	*comfortable*
The oscilloscope measured the voltage.	*marginal*
The oscilloscope calculated the voltage.	*uncomfortable*

As long as objects perform the actions that they're endowed to perform, the reader will not trip.

Some passive voice arises in scientific writing because scientists and engineers cling to the misconception that they can never use the first person ("I" or "we"). Well, in his writings, Einstein occasionally used the first person. He was not only a great scientist, but a great scientific writer. Feynman also used the first person on occasion, as did Curie, Darwin, Lyell, and Freud. As long as the emphasis remains on your work and not you, there is nothing wrong with judicious use of the first person.

Also, avoiding the first person at every turn leads to unnatural wording:

> In this paper, the authors assumed that all collisions were elastic.

The phrase "the authors assumed that" is silly. The reader can see the authors' names on the paper. Using the word "authors" instead of "we" suggests that the paper had a ghostwriter. Other unnatural phrases that arise from avoiding the first person include

> It was determined that...
>
> It was decided that...
>
> It was attributed to...

These phrases suggest that there was some absolute force—the "It" force—"determining," "deciding," or "attributing." The "It" force suggests that your conjectures are more than conjectures and that some being greater than mortal man performed the work. In many cases where you make an assumption or decision, the first person is not only natural, but almost obligatory. Consider an example in which the writers should have used the first person, but did not:

> In that an effort to identify a specific control circuit responsible for the failure of the gear box was unsuccessful, it was determined appropriate to resurvey the collector field for torque tube damage.

In this example, by avoiding the first person, the engineers shirked responsibility for not finding what caused the gear box failure and for authorizing another survey of the field. Appropriate use of the first person would have been as follows:

> Because we couldn't locate the control circuit responsible for the gear box failure, we surveyed the collector field again for damage to the torque tubes.

When using the first person, how do you keep the emphasis on the work and not the person(s) doing the work? First, you should reserve the use of the first person for those occasional situations in which your role in the work is important—for instance, when you make an assumption. Second, you should avoid placing the first person (either "I" or "we") as the beginning word of a sentence because that position receives heavy emphasis. Instead, have the first person follow an introductory adverb, infinitive phrase, or dependent clause.

Even though you use the first person judiciously and keep the emphasis on the work and not on yourself, you will inevitably run across managers or editors who will delete the first person from your drafts. In such cases, I have found the best thing to do is to disagree politely, perhaps even raise a rebuttal based on reason. However, if the manager or editor stands by the edit, you should accept the edit as a format constraint for that document. You will find, though, that by continuing to use the first person judiciously on subsequent documents, you will wear down the manager's or editor's resentment of the first person. For one thing, the manager or editor will find it difficult to edit out appropriate use of the first person without fattening the sentences. Second, once the manager or editor becomes accustomed to hearing judicious use of the first person in a document, the manager or editor will realize the natural sound that the first person gives the writing.

Chapter 7

Language: Being Familiar

The whole of science is nothing more than a refinement of everyday thinking.

—Albert Einstein

To inform your audience, you have to use language that your audience understands. When you have an audience outside your area of expertise, it is often difficult to find familiar language. Why? Scientists and engineers work on particular crooks of particular ridges of particular mountains, and each mountain has its own special set of terms. These terms may be words or abbreviations:

Plasma Physics:	glow discharge	MHD
Hemodynamics:	immunodeficiency	B cells
Spectroscopy:	gas chromatography	GC/MS

In a scientific document the writer, not the reader, bears the responsibility of bridging the language gap. When you write a scientific document, you should ask yourself whether your terms are familiar. If your terms are not familiar, then you should either write around them or define them. Likewise, if you introduce an unfamiliar concept to your audience, then you should find a way, such as an example or analogy, to explain that concept.

Avoiding Unfamiliar Terms

One of the simplest, yet most overlooked, ways to handle unfamiliar terms is to avoid them. This advice is especially pertinent when the unfamiliar term is *jargon*, a vocabulary particular to a place of work. Jargon may be either abbreviation or slang. Don't assume that jargon is inherently bad. For communication within a place of work, jargon can concisely identify experiments and buildings. Let's say that you work at a laboratory called the Pulsed Power Facility. When you write internal memos and reports, you might use the facility's jargon:

PPF: Pulsed Power Facility

Jarva: a neodymium-glass laser in the facility

For readers who work at the facility, this jargon poses no problem. However, when you write a journal article or an external report, this jargon alienates readers. In handling a piece of jargon, such as "Jarva," you can simply avoid the term. In the document, you would just use the "neodymium-glass laser" the first time, and the "laser" thereafter. For readers outside the facility, what difference does it make if your laser is named "Jarva" or "Olivia"?

The Department of Defense has become a master at creating jargon. In fact, the Department publishes a huge book each year that defines its current jargon. In the book, you'll find definitions for "Jolly Green Giant" (military transport helicopter), "Pony Express" (radar optics intelligence), and "Jack Frost" (winter military exercise). Some jargon has multiple meanings. For instance, the abbreviation ABC stands for one of the following four things: advanced blade concept; Argentina, Brazil, and Chile (the ABC countries); American, British, and Canadian; or atomic, biological, and chemical.

As a scientist or engineer, you want to challenge jargon and use it only when it makes the reading more efficient for your readers. Many scientists and engineers unfortunately go out of their way to include every piece of jargon in their work, even when they are writing to readers outside their place of work:

> For the first year, the links with SDPC and the HAC were not connected, and all required OCS input data were artificially loaded. Thus, CATCH22 and MERWIN were not available.

This paragraph reads like a cryptogram. Whenever precision allows, you should use common words, not unfamiliar jargon:

> Because some of the links in the computer system were not connected the first year, we could not run all the software codes.

Jargon not only alienates; it often misleads:

> The 17 percent efficiency of the water/steam system was a showstopper.

Someone not familiar with the word "showstopper" might interpret 17 percent as a favorable efficiency, so favorable that the show was interrupted by applause. In this particular sentence, however, the engineer meant just the opposite: The efficiency was so poor that the system was disregarded and the project was cancelled.

Defining Unfamiliar Terms

When you cannot write around unfamiliar terms, then you have to define them. Notice that unfamiliar terms include both usual words, such as "bremsstrahlung," that are particular to an area of work, and common words, such as "receiver," that have unusual meanings in an area of work.

What is the best way to define a term? If a definition is short, you can include it within the sentence:

> In a central receiver system, a field of sun-tracking mirrors, or *heliostats*, focuses light onto a solar-paneled boiler, or *receiver*, mounted on top of a tall tower.
>
> The thermal oil is a mixture of many *aliphatic hydrocarbons* (paraffins).

When the definition is complex or unusual, you should expand your definition to a sentence or two.

> *Bremsstrahlung* is the radiation emitted by a charged particle that is accelerated in the Coulomb force field of a nucleus. The term originates from the German words *Bremse* (brake) and *Strahlung* (radiation).
>
> *Atherosclerosis* is a disease in which fatty substances line the inner walls of the arteries. If these deposits plug an artery, then a heart attack, a stroke, or gangrene may occur.

In your definitions, select words that are familiar to your readers; otherwise, your definitions won't inform.

> *Serum lipids* are cholesterols and triglycerides transported in the blood.

If your readers do not know the meaning of "serum lipids," chances are they won't know the meaning of "triglycerides," either. Here, you should either replace "triglycerides" with a familiar word, or define "triglycerides" within your definition of "serum lipids."

When formally defining a noun term, you should begin the definition with a familiar noun that identifies the class to which the term belongs. You should then provide enough information to separate that term from all other terms in the class.

> *Retina* is light-sensitive tissue, found at the back of the eye, that converts light impulses into nerve impulses.

Here, the noun phrase "light-sensitive tissue" establishes the class to which the term "retina" belongs, and the additional information separates retina from other types of light-sensitive tissues. Notice that the following definitions are incomplete:

Cholesterol is present in body fluids and animals.

Triglycerides are fats.

The first definition needs a noun such as "fat" to classify "cholesterol," and the second definition needs additional information to separate "triglycerides" from other kinds of fats.

Defining abbreviations is a little different. If the abbreviation occurs only a couple of times in the document, then you should avoid the abbreviation and write out the expression. For instance, if a document uses the abbreviation "CRS" (central receiver system) fewer than three times, do not bother to include it. Instead, use "central receiver system." If, however, the abbreviation occurs several times, then you should define the abbreviation the first time it's used:

> In a central receiver system (CRS), a field of solar mirrors focuses sunlight onto a central boiler or receiver. An example of a CRS is the Solar One Power Plant located near Barstow, California. Another example of a CRS is...

If a term appears only once or twice in the beginning of a long document, but several times in a later section, you would make things easier on your audience if you would use the full expression in the earlier references and define the abbreviation in the later section when the abbreviation will be used. Too often, scientists and engineers define an abbreviation on one page and then do not use that abbreviation until several pages later. How do these scientists and engineers expect their readers to remember an abbreviation from several pages before? What happens is that the reader forgets the definition and then has to scan back through the document looking for the abbreviation's definition—an exercise that does not make for efficient reading.

Note that some abbreviations such as DNA or AIDS are more common than the terms they abbreviate. When an abbreviation is more commonly known than the corresponding term, you should define the abbreviation somewhere in the document for completeness' sake, but treat the abbreviation as the familiar term in titles, summaries, and the main text.

Incorporating Examples and Analogies

Two of the best tools for explaining unfamiliar concepts are examples and analogies. Scientists and engineers know the value of examples—the favorite mathematical textbooks were the ones with the most sample problems. In your writing, you have many occasions to use examples. For instance, whenever you make general statements, you should anchor these statements with examples. Don't leave your readers clutching to generalities:

> Since the design of the Solar One Power Plant, significant advances have occurred in solar energy technology.

Unless anchored with an example, this statement will soon be forgotten because it relies on the general phrase "significant advances." You need a specific example that readers will remember:

> Since the design of the Solar One Power Plant, significant advances have occurred in solar energy technology. For example, experimental tests have shown that using molten salt, rather than water, as the heat transfer fluid could increase overall system efficiency from 17 percent to 25 percent.

Scientists and engineers often miss opportunities to enrich their writing with examples. Suppose, for example, that you had to justify the importance of supercomputers to a non-technical audience. Many scientists and engineers would just give the amount of storage or the number of arithmetic operations per second. One computer scientist [Wilson, 1985], though, took a more careful approach.

In his proposal to the Department of Defense, he created an example to show how powerful supercomputers are:

> By the late Middle Ages, cities throughout Europe were building Gothic cathedrals. In this effort, citizens dedicated their entire lives to the construction of a single cathedral, often in competition with other cities for the highest or widest building. The only way, however, that architects could test a new design was to build a cathedral, a process that took over forty years. Unfortunately, many cathedrals caved in during or after construction because the designers had no way other than trial and error to test their designs. What took forty years to test in the Middle Ages could have been done in minutes on a supercomputer.

The author backed up his example with a figure showing a supercomputer design of a cathedral with a height of 32 meters that proved successful. Next to this design was a supercomputer design of a cathedral with a height of 41 meters that failed. After the proposal was published and distributed, the author flew to Washington to give follow-up presentations. The author claimed that every proposal reviewer he met remarked about the cathedral example. Examples are what people remember.

Besides examples, analogies are also valuable at conveying complex ideas or numbers. Analogies compare obscure thoughts to familiar ones. Einstein used them generously in his writing:

> I stand at the window of a railway carriage which is travelling uniformly, and drop a stone on the embankment, without throwing it. Then, disregarding the influence of air resistance, I see the stone descend in a straight line. A pedestrian who observes the misdeed from the footpath notices that the stone falls to earth in a parabolic curve. I now ask: Do the "positions" traversed by the stone lie "in reality" on a straight line or on a parabola? [Einstein, 1945]

Einstein's analogy is so much more alive than the abstract question: Where do positions of an object lie in reality? Unfortunately, many scientific documents are devoid of analogies. Analogies are tools to help readers. Analogies demand imagination and creativity. The best scientists and engineers use them. Analogies not only help readers understand concepts; they also provide insights into how writers think. Consider an analogy by Sigmund Freud [1933]:

> One might compare the relation of the ego to the id with that between a rider and his horse. The horse provides the locomotive energy, and the rider has the prerogative of determining the goal and of guiding the movements of his powerful mount towards it. But all too often in the relations between the ego and the id we find a picture of the less ideal situation in which the rider is obliged to guide his horse in the direction in which it itself wants to go.

Besides providing analogies to explain ideas, you will find opportunities to provide analogies to show the significance of numerical findings:

> In the brightness tests, the maximum retinal irradiance was less than 0.064 w/cm^2, a brightness about that of a household light bulb.
>
> The Hanford Nuclear Reservation contains 570 square miles, which is about half the size of Rhode Island.

Numerical analogies make your writing unique. Consider how Feynman [1964] gives significance to the magnitude of electrical forces.

> If you were standing at arm's length from someone and you had one percent more electrons than protons, the repelling force would be incredible. How great? Enough to lift the Empire State Building? No. To lift Mount Everest? No. The repulsion would be enough to lift a weight equal to that of the entire earth.

References

Einstein, Albert, *Relativity: the Special and General Theory*, trans. by R. W. Lawson (New York: Crown Publishers, 1945), p. 9.

Einstein, Albert, "Physik und Realität," trans. by J. Piccard, *Journal of the Franklin Institute*, vol. 221 (March 1936), p. 313. Original text: *"Alle Wissenschaft ist nur eine Verfeinerung des Denkens des Alltags."*

Feyman, Richard P., Robert B. Leighton, and Matthew Sands, *The Feynman Lectures on Physics*, vol. II (Reading, MA: Addison-Wesley Publishing Co., 1964), p. 1.

Freud, Sigmund, "The Anatomy of the Mental Personality," Lecture 31, *New Introductory Lectures on Psychoanalysis*, trans. by W. J. H. Sprott (New York: Norton, 1933), p. 108.

Wilson, William D., and Robert Gallagher, "The Need for Supercomputers in Nuclear Weapon Design" (Livermore, CA: Sandia National Laboratories, 1985), pp. 10–11.

Chapter 8

Language: Being Concise

Vigorous writing is concise. A sentence should have no unnecessary words, a paragraph no unnecessary sentences, for the same reason that a drawing should have no unnecessary lines and a machine no unnecessary parts. This requires not that the writer make every sentence short, or that he avoid detail and treat his subject only in outline, but that every word tell.

—William Strunk

Conciseness follows from pursuing two other language goals: being clear and being forthright. When you make your writing clear and forthright, you also tighten it. Ridding sentences of pretentious diction such as "utilize" and "facilitate" leaves the more concise verbs "use" and "make." Ridding sentences of abstract nouns such as "factor" and "nature" cuts the fat prepositional phrases that accompany those nouns. This chapter discusses four ways to cut fat in scientific writing: eliminating redundancies, eliminating writing zeroes, reducing sentences to simplest form, and cutting bureaucratic waste.

Eliminating Redundancies

Redundancies are needless repetitions of words within a sentence. Redundancies either repeat the meaning of an earlier expression or else make points implicit in what has been stated. Adjectives are often redundant:

> The aluminum metal cathode became pitted during the glow discharge.

After "aluminum," the adjective "metal" is redundant. Like adjectives, adverbs are often redundant.

> The use of gaseous insulation is becoming increasingly more widespread.

The verb phrase "is becoming increasingly more widespread" is doubly redundant. Revision gives

> Scientists are using gases more as insulators.

Below are some common redundancies in scientific writing. The words in parentheses can be deleted.

(already) existing	introduced (a new)
(alternative) choices	mix (together)
at (the) present (time)	never (before)
(basic) fundamentals	none (at all)
(completely) eliminate	now (at this time)
(continue to) remain	period (of time)
(currently) being	(private) industry
(currently) underway	(separate) entities
(empty) space	start (out)
had done (previously)	(still) persists

There are many more redundancies besides the ones listed here. Because everyone writes redundancies in early drafts, you have to catch redundancies in your editing. An effective way is to read with the sole intention of cutting words—no additions allowed.

Eliminating Writing Zeroes

A second aspect of being concise is to eliminate writing zeroes. Certain phrases have no meaning at all. In your writing, these phrases are zeroes: voids that offer no information to your readers.

> It is interesting to note that over 90 incidents of satellite fragmentations have produced over 36,000 kilograms of space debris.

The phrase "it is interesting to note that" is a zero. If the detail isn't interesting, then the writer shouldn't include it, and if the detail is more interesting than other details, then the writer should find a stronger way to highlight it, perhaps by placing the sentence at the beginning or end of a paragraph.

Sometimes, writing zeroes raise undesirable questions. Consider the following example:

> The requirements to be met for the detection system of plastic explosives include a detection rate of at least 95 percent and a false alarm rate of less than 5 percent.

The writing zero "to be met" is dangerously superfluous. It implies that there are requirements which will not be met.

Some other writing zeroes are

as a matter of fact	it should be pointed out that
I might add that	the course of
it is noteworthy that	the fact that
it is significant that	the presence of

Although these deletions appear small, they are important. Consider the following example:

> Vibration measurements made in the course of the missile's flight test program were complicated by the presence of intense high-frequency excitation of the vehicle shell structure during the re-entry phase of the flight. *(33 words)*

Eliminating the zeroes "in the course of" and "the presence of" not only saves a few seconds of reading time, but also invigorates the writing. Notice also the redundancies "vehicle shell (structure)" and "re-entry (phase of the flight)."

> Vibration measurements made during the missile's flight were complicated by intense high-frequency excitation of the vehicle shell during re-entry. *(19 words)*

This revision is about half the length of the original. More important, the audience can comprehend this revision much faster. Concise writing is crisp writing.

Reducing Sentences to Simplest Forms

A third way to be concise in your writing is to reduce sentences to their simplest forms. Just as you reduce mathematical equations to their simplest forms for easier comprehension, so too should you reduce sentences to their simplest forms. Reducing sentences to their simplest forms does not mean limiting yourself to writing only simple sentences (a sentence structure with only one independent clause). Rather, reducing sentences to their simplest forms means using only the necessary words in whatever sentence structure you choose, be it simple, compound, or complex.

Many reductions occur within phrases. Consider the following examples:

Fat Phrase	Reduction
at this point in time	now
at that point in time	then
has the ability to	can
has the potential to	can
in light of the fact that	because
in the event that	if
in the vicinity of	near
owing to the fact that	because
the question as to whether	whether
there is no doubt but that	no doubt

Other reductions for sentences are not so straightforward. How do you find these reductions? In reducing mathematical equations, you look for signals such as the same variable appearing on both sides of the equation. In reducing sentences, you should look for signals as well.

One signal is the overuse of adjectives. An adjective is a word that modifies a noun. Often you require adjectives because you don't have a noun specific enough to stand for the person, place, or thing you are describing. Nonetheless, you should challenge adjectives and weed out those that don't serve a purpose in your sentences.

> The objective of our work is to obtain data that can be used in conjunction with a comprehensive chemical kinetics modeling study to generate a detailed understanding of the fundamental chemical processes that lead to engine knock. *(37 words)*

The adjectives "comprehensive," "detailed," and "fundamental" don't add anything here. Cutting the needless adjectives as well as the redundancies and writing zeroes, we get

> Our goal is to obtain experimental data that can be used with a chemical kinetics model to explain the chemical processes that lead to engine knock. *(26 words)*

Having a lot of adverbs is another signal that your sentence could be reduced. An adverb is a word that modifies a verb, an adjective, or another adverb. Sometimes, adverbs serve important purposes in sentences. All too often, though, adverbs are excess baggage. For instance, because some adjectives are absolutes, using an adverb to modify these adjectives is illogical. An example of such an adjective is "unique." "Unique" means unlike anything else. For that reason, something is either unique or not unique—nothing in between. Therefore, the phrase "somewhat unique" makes no sense; it falls in the same category as "somewhat perfect."

Other times, using an adverb such as "very" cripples the word that the adverb modifies. For instance, if you describe a result as "very important" in the beginning of a document, you've essentially sapped the power from the word "important" for the rest of the document. From then on, any "important result" seems commonplace. Illogical and crippling uses of adverbs clutter sentences and reduce the energy of the writing:

> The achievement of success in these advanced technologies depends very heavily on a rather detailed understanding of the complex processes that govern the velocities in the unburned gases prior to combustion. *(31 words)*

"Depends" is a strong verb. However, coupling "depends" with "very heavily" cripples its meaning. The same is true for the noun "understanding," which is coupled with "rather detailed." Cutting these needless adverbs and the baggage that they bring to the sentence not only shortens the writing, but also increases the writing's energy:

> The success of these advanced technologies depends on understanding the velocities in the unburned gases prior to combustion. *(18 words)*

Nouns containing verbs are a third signal that your sentences could be reduced. Turning verbs into nouns attracts needless adjectives and adverbs.

> The establishment of a well proven, well documented, rational methodology for making precise velocity measurements in the unburned gas has been realized and is being used extensively to aid in the development of numerical models, which in turn are used in the design of advanced piston engines. *(45 words)*

"Establishment," "measurement," and "development" hide verbs that could unlock the fat hidden in this sentence. Much of this fat occurs before the noun "methodology." If you pull out a five-dollar noun such as "methodology," it should at least pay its own way through the sentence. This writer tossed "methodology" into a pile of unnecessary adjectives and adverbs. Revision gives

> We have found a method to measure velocity in the unburned gas, and with this method, we are developing numerical models that will help design advanced piston engines. *(27 words)*

Finally, needlessly passive verbs represent another signal that your sentences could be reduced to a simpler form. Much fat in scientific writing arises from needless use of passive voice:

> It was then concluded that a second complete solar mirror field corrosion survey should be conducted in July to determine whether the tenfold annual corrosion rate projection was valid and to allow determination as to whether subsequent corrective measures would be effective in retarding corrosion propagation. *(85 syllables)*

Revision gives

> To see whether the corrosion rate would increase tenfold as projected and to see whether stowing the mirrors in a vertical position would slow the rate, we decided to survey the solar mirror field a second time in July. *(61 syllables)*

Although this revision did not significantly reduce the number of words in the sentence, it did significantly reduce the number of syllables. Syllable count affects reading efficiency as much as word count. In reducing the number of syllables, we have replaced vague and needlessly complex phrases such as "subsequent corrective measures" with precise and clear phrases such as "stowing the mirrors."

Eliminating Bureaucratic Waste

So far, this chapter has looked at being concise from the phrase and sentence levels. A broader perspective comes from examining needless paragraphs and sections. Much management writing is wasteful. In this kind of writing, empty nouns such as "target," "parameter," and "development" fill the page without examples to anchor the meanings of those nouns. Before cutting through the deadwood at the phrase and sentence levels of management writing, you should consider who the audience is for the document and what the audience wants to learn from the document. That thinking will often help delete entire paragraphs and sections.

Consider this preface written by the Department of Energy [1985] for the reports in one of its solar energy programs. The audience for the preface was varied: solar engineers, contract engineers, congressmen, and the general public:

> The research and development (R & D) described in this document was conducted within the U. S. Department of Energy's (DOE) Solar Thermal Technology Program. The goal of the Solar Thermal Technology Program is to advance the engineering and scientific understanding of solar thermal technology, and to establish the technology base from which private industry can develop solar thermal power production options for introduction into the competitive energy market.
>
> The Solar Thermal Technology Program is directing efforts to advance and improve promising system concepts through the research and development of solar thermal materials, components, and subsystems, and the testing and performance evaluation of subsystems and systems. These efforts are carried out through the technical direction of DOE and its network of national laboratories who work with private industry. Together they have established a comprehensive, goal-directed program to improve performance and provide technically proven options for eventual incorporation into the Nation's energy supply.
>
> To be successful in contributing to an adequate national energy supply at reasonable cost, solar thermal energy must eventually be economically competitive with a variety of energy sources. Components and system-level performance targets have been developed as quantitative program goals. The performance targets are used in planning research and development activities, measuring progress, assessing alternative technology options, and making optimal component

> developments. These targets will be pursued vigorously to insure a successful program.

In this preface, the Department of Energy did not anchor its nouns. For instance, why go into detail about the "targets" or "goals" in the third paragraph if you don't let the readers know what those targets and goals are? The readers who understood this preface were not the readers of the program's solar reports. The readers who understood this preface were the Department of Energy members who wrote the preface. In other words, the Department of Energy wrote this preface to itself. Revision with consideration of the audience gives

> The research described in this report was conducted within the U. S. Department of Energy's Solar Thermal Technology Program. This program directs efforts to incorporate technically proven and economically competitive solar thermal options into our nation's energy supply. These efforts are carried out through a network of national laboratories that work with industry.

Although this revision is less than half the length of the original, it has more than twice the power. Fat writing is lethargic writing. Concise writing moves.

References

Department of Energy, *Solar Thermal Technology Annual Evaluation Report* (Golden, CO: Solar Energy Research Institute, 1985), p. ii.

Strunk, W., Jr., and E. B. White, *The Elements of Style*, 3rd ed. (New York: Macmillan Publishing Co., 1979), p. 23.

Chapter 9

Language: Being Fluid

The greatest possible merit of style is, of course, to make the words absolutely disappear into the thought.

—Nathaniel Hawthorne

Many scientists and engineers mistakenly believe that scientific writing must be dull. Unfortunately, their writing reflects that misconception. It is dull, needlessly dull. Many scientists and engineers undercut the purpose of scientific writing—to inform—with sentences and paragraphs that drag and with discontinuities in language that trip readers. Scientists and engineers study the most fascinating subjects in the world: solar systems, animal and plant kingdoms, the inner workings of atoms and nuclei. Why then do so many scientists and engineers resign themselves to prose styles that are sluggish, that are without life?

No one expects you to write with the grace of John Cheever or Joyce Carol Oates, but you do have to inform, and if you're going to inform, you can't bore readers with sluggish language—language without variety, language full of discontinuities. Your language should be precise and clear. It should be anchored in the familiar. It should be forthright and concise, and it should be fluid. Fluid language is the lubricant that makes your writing inform. Although the constraints of scientific writing somewhat limit its stylistic variation, scientific language can still be fluid. It can be energetic, sometimes exciting.

Varying Sentence Rhythms

Many scientific papers read slowly because the sentences have no rhythmic variety. In any type of writing, be it a poem or lab report, there are rhythms. Rhythms help determine the energy of the writing. Writing that uses the same rhythms over and over is dull reading. Imagine a piano piece with only two or three different notes—not particularly exciting. But that is the way many scientists and engineers write: two or three sentence patterns repeated again and again. The sentences begin the same way, the sentences have the same length, and the sentences have the same arrangement of nouns, verbs, and phrases. The result is stagnation.

> Mount St. Helens erupted on May 18, 1980. A cloud of hot rock and gas surged northward from its collapsing slope. The cloud devastated more than 500 square kilometers of forests and lakes. The effects of Mount St. Helens were well documented with geophysical instruments. The origin of the eruption is not well understood. Volcanic explosions are driven by a rapid expansion of steam. Some scientists believe the steam comes from groundwater heated by magma. Other scientists believe the steam comes from water originally dissolved in the magma. We have to understand the source of steam in volcanic eruptions. We have to determine how much water the magma contains.

The subject matter is interesting, but the prose is tiresome and the rhythms are monotonous.

How do you vary rhythm? There are many things that affect rhythm: the way sentences begin, the way sentences end, the position of the subject and verb. Some of these effects are stronger than others. These stronger effects include the way sentences begin, the lengths of sentences, the structure of sentences, and the lengths of paragraphs.

Varying Sentence Openers. Many ways exist to begin a sentence:

Subject-Verb	*Mount St. Helens erupted* on May 18, 1980.
Prepositional Phrase	*Within minutes,* the cloud devastated more than 500 square kilometers of forests and lakes.
Transition Words	*Recently,* debate has arisen over the source for the steam.
Introductory Clause	*Although the effects of the eruption were well documented,* the origin is not well understood.
Infinitive Phrase	*To understand the source of steam in volcanic eruptions,* we have to determine how much water the magma contains.
Participial Phrase	*Its slope collapsing,* the mountain emitted a cloud of hot rock and gas.
Verb (Question)	*Is* it groundwater heated by magma or water originally dissolved in the magma itself?

The subject-verb opener is a staple of scientific writing. Because subject-verb openers are the most direct way to state details, they are valuable for opening sentences to sections and subsections and for stating important results. As shown in the earlier paragraph on volcanoes, though, writing that relies solely on subject-verb openers is stagnant.

Another common sentence opener, the prepositional phrase, is important for making connections between sentences. Preposition phrases often include details about time or position. If you place such a prepositional phrase at the beginning of the sentence and that detail echoes a detail from the previous sentence, then you have made a connection.

Beginning a sentence with a transition word such as "moreover," "however," or "therefore" also makes a nice transition because after reading only one word of a sentence, the audience knows the relationship of that sentence to the preceding one. Now, your English teacher may have told you not to begin a sentence with "however." At one time, beginning a sentence with "however" was not considered proper usage. However, what's considered proper usage varies over time, and beginning a sentence with "however" is now accepted. For more discussion about accepted usage in scientific writing, see Appendix B.

Infinitive phrases and dependent clauses (particularly those clauses that answer questions about why, when, and where) are also important in scientific writing. Because sentences beginning with infinitive phrases and dependent clauses let the audience know early on that the sentence has two parts, you can generally write longer sentences with these types of beginnings. As stated in Chapter 5, longer sentences are fine as long as the sentence keeps to one main idea and the audience realizes early on that the sentence will be long.

Although beginning a sentence with a participial phrase dramatically changes the rhythm, the practice is not as common as it once was. The participial phrase opener, which was common in Latin, is rarely used today in spoken English. For example, think of the last time that you constructed a sentence in your speech that used a participial phrase opener.

> Walking into the laboratory, I turned on the light.
>
> Feeding the elephant, I noticed a tumor.

Because the construction is so uncommon in speech, it appears awkward in writing. Granted, writing is not limited to the sentence rhythms that we speak, but readers have internal ears that have been tuned by speech. Reading a sentence rhythm not echoed in speech stands out, sometimes undesirably so. Also, long participial phrases often frustrate readers because they have to wait so long to find out what the phrase modifies that they forget what the phrase contains.

> Originating at 500 meters beneath a bulge on the north face and attaining a velocity of 175 meters per second, the volcanic blast was captured by photography.

This sentence does not efficiently communicate because the audience does not fully appreciate the details in the long participle phrase until after seeing what the subject is. In other words, the sentence has to be read twice. If you use participial phrase openers, keep them short and infrequent.

The direct question is a sentence structure whose rhythm is the opposite of a subject-verb opener. Once you have presented your readers with a set of details, questions are wonderful for showing readers what perspective to take in examining those details. Questions are also wonderful for making the reader an active participant in the document. When should you use questions? Perhaps an easier question to address is, When should you not use a question? Questions at the beginnings of documents are risky. Because so many college freshman essays have begun with questions, starting any type of document with a question risks cliché. Also, ending a document with a question carries the risk of frustrating readers. At the end of a document, readers expect closure. Unanswered questions do not provide closure. A question seems to work best in the middle of a discussion when a set of details has been introduced. The question serves to orient the reader by having the reader look at those details from a single perspective.

So far, this subsection has presented some common

ways to begin sentences. There are certainly others, such as *appositives*, *gerundial phrases*, and *introductory series*. No matter how many sentence openers are in your repertoire, you are faced with the question of how to vary those openers to achieve the best rhythm. That question is difficult, if not impossible, to answer. As an experiment, consider having all seven openers from this subsection in a single paragraph:

> Mount St. Helens erupted on May 18, 1980. Its slope collapsing, the mountain emitted a cloud of hot rock and gas. Within minutes, the cloud devastated more than 500 square kilometers of forests and lakes. Although the effects of the eruption were well documented, the origin is not well understood. Volcanic explosions are driven by a rapid expansion of steam. Recently, debate has arisen over the source for the steam. Is it groundwater heated by the magma or water originally dissolved in the magma itself? To understand the source of steam in volcanic explosions, we have to determine how much water the magma contains. [Eichelberger, 1983]

The rhythms of this paragraph are pleasing. However, don't assume that you must use these seven sentence openers in this or any other set pattern. Strong writing contains no magic formulas for sentence openers or any other aspect of style. Rather, the way you vary sentence openers helps determine your individual style. Just remember that failure to vary sentence openers will stagnate your prose and exhaust your readers.

Varying Sentence Lengths. Another important way to vary sentence rhythm is to vary sentence length. As stated in Chapter 5, you would like to keep your average sentence length below twenty words. Although the average length should be below twenty words, the sentence lengths can have a wide range. Not having a wide range stagnates the writing because the stop signs (the periods at the ends of sentences) occur at regular intervals.

> On the morning of May 18, a strong earthquake shook Mount St. Helens, causing the volcano's cracked and steepened north side to slide away. (24) Photographs taken during these early seconds, together with other information, showed that the blast originated 500 meters beneath a bulge on the north face. (25) Photographs and time of destruction of a seismic station established the velocity of the blast to be about 175 meters per second. (22) Of significance, the volume of new magmatic material ejected in the blast (about 0.1 km^3) equals the volume of the bulge. (22)

This paragraph is readable, but the sentence length distribution is narrow (22–25 words), making the writing monotonous.

What distribution of sentence lengths should you use? The answer here is the same answer as for what distribution of sentence openers you should use—no one knows. However, here are some general guidelines:

1. Try to keep your average length in the teens.
2. Change your sentence lengths often, at least every two or three sentences.
3. Occasionally use a short or long sentence.

How short can you make your sentences? You can make your sentences as short as you'd like. An occasional four- or five-word sentence can accent an important result, particularly if the sentence follows a rather long sentence. Don't think that using an occasional short sentence will make your writing appear simplistic. George F. Will, for instance, scatters short sentences throughout his essays, and although many people disagree with his political views, few people accuse him of being simplistic. Be careful with short sentences, though. Stacking short sentences will make your writing choppy. How long can a sentence be? In principle, as long as you want, as long as you maintain clarity.

> On the morning of May 18, a strong earthquake shook Mount St. Helens, causing the volcano's cracked and steepened north side to slide away. (24) Photographs showed that the blast

began beneath a bulge on the north face. (13) These photographs, when coupled with other information, established the depth of the blast's origin to be 500 meters and the velocity of the blast to be 175 meters per second. (31) The measured volume of the ejected magmatic material was about 0.1 km^3. (11) This volume equals the volume of the bulge. (8)

Varying Sentence Structure. A third way to vary sentence rhythm is to vary the sentence structure. Three common sentence structures are simple sentences, compound sentences, and complex sentences. A simple sentence is a sentence with a single independent clause:

> *Lava* from a nonexplosive eruption ordinarily *contains* only 0.2 percent water.

In this simple sentence, the subject is "Lava" and the verb is "contains." A compound sentence is a sentence that contains two or more independent clauses. These clauses are usually joined by a conjunction such as "and" or "but."

> Precursor activity to the eruption began on March 20, 1980, *and* many times during the next two months the mountain shook for minutes.

The conjunction in the above compound sentence is "and." A complex sentence is a sentence that contains an independent clause joined with one or more dependent clauses (note that the word "complex" here does not have the same meaning that it did in Chapter 5).

> *Although the amount of devastation caused by the May 18 blast was a surprise,* the eruption itself had been expected for weeks.

The dependent clause in the above complex sentence is in italics.

How does varying sentence structure vary rhythm? By varying sentence structure, you ensure that you vary the location and number of the subjects and verbs in your sentences. Subjects and verbs receive heavy accents in sentences. When you vary the location and number of those accents within your sentences, you vary your rhythm. There are other kinds of sentence structures, such as compound-complex, but don't worry about them. It's not important that you know the names of all the sentence structures or even the names mentioned here. What is important is that your sentence structures don't become stagnant. As with varying sentence lengths and sentence openers, no set distributions exist for varying sentence structures. The advice is simple: If your sentence structures begin to look the same, vary them.

Varying Paragraph Lengths. A paragraph is a group of connected sentences that conveys a group of connected ideas. When a sentence ends, there is a period and then a space before the next sentence begins. This space break is analogous to a stop sign. When a paragraph ends, though, there is a break for the rest of the line, and then the writing starts after an indent on the next line. This break is much longer and is analogous to a traffic light.

Paragraph length is not measured by the number of sentences or even the number of words. Rather, paragraph length is measured by the number of lines that the reader sees on the page. When paragraphs are consistently short (one to seven lines), then the reader sees too many traffic lights and the reading after a page or so becomes tiresome because it is stop and go, stop and go. When the paragraphs are consistently long (fourteen or more lines), then the reader tires in another way. The reader must digest too much information without a sufficient number of rest stops. Extremely long paragraphs intimidate the audience before the audience has even begun reading them because the audience can see where the visual breaks lie on the page. To avoid that intimidation, many writers go no more than two-thirds down a page without either a new paragraph or a white space break (say, for an equation) within the paragraph.

In general, you would like to have most of your paragraphs fall in the middle category, between seven and fourteen lines long, with an occasional short paragraph (one to seven lines) and an occasional long paragraph (fourteen or more lines). This pattern is pleasing to an audience, not only when the audience reads through the document, but also when the audience scans the pages.

Note that paragraph length depends on the format. When you are writing for a three-column format, such as a newsletter, your paragraphs tend to be shorter in the number of words and number of sentences than when you are writing in a single-column format, such as a formal report. Also, note that the situation affects paragraph length. Some short documents, such as correspondence, tend to have shorter average paragraphs (four to six lines) than other documents, such as journal articles and formal reports, where the average length is between seven and fourteen lines.

Eliminating Discontinuities

Being fluid means making smooth transitions between sentences and paragraphs. Discontinuities arise in language for a variety of reasons: poor transitions between ideas, needlessly complex typography, and poor transitions between text and equations. This section presents ways to bridge those discontinuities.

Making Transitions Between Ideas. Some discontinuities arise from the writer waiting too long to link ideas in sentences:

> The Cascade Range, with its prominent chain of towering cones, is not the only threatening volcanic region in the western United States. Many people who live in the eastern Sierra Nevada community of Mammoth Lakes, California, may have been unaware until recently that their scattered hills and ridges have a remarkably recent volcanic origin as well.

Between these two sentences, the tie does not occur until the last few words of the second sentence. The geologist keeps her readers in limbo for over thirty words before making a connection. That's too long. Use connective words early in sentences to make strong transitions with the previous sentences:

> The Cascade Range, with its prominent chain of towering cones, is not the only threatening volcanic region in the western United States. The Mammoth Lakes area of the Sierra Nevada also has a recent volcanic origin.

The word "also" ties the second sentence to the first. Words such as "also" are transitional words. Transitional words signal readers one of three things: (1) the movement of ideas will continue in the same direction, (2) the movement of ideas will pause, or (3) the movement of ideas will reverse direction.

Continuation	Pause	Reversal
also	for instance	however
moreover	for example	on the other hand
first...second...third	in other words	conversely

Other causes for poor transitions between ideas are gaps in logic—places where key details have been omitted. The writer doesn't trip over the gaps because the writer knows the key details between the ideas; however, the readers trip. In the Cascade Range example, the geologist assumed her readers knew that the Sierra Nevada was a separate mountain range from the Cascade Range. This assumption depended on the audience. If the audience was diverse, the example would better read

> The Cascade Range, with its prominent chain of towering cones, is not the only threatening volcanic region in the western United States. Farther south, the Mammoth Lakes area of the Sierra Nevada Range also has a recent volcanic origin.

The word "range" after "Sierra Nevada" and the phrase "Farther south" strengthen the connection between the two sentences.

Eliminating Needless Complex Typography. Run your eye over a page of writing from a science or engineering journal. Now run your eye over a page of writing in the newspaper. Which page appears the more intimidating? The page from the journal probably does because of the typography inherent to science and engineering. Abbreviations, numerals, and strings of capital letters bring discontinuities into scientific writing. Although their use is sometimes unavoidable, many scientists and engineers strew these typographical discontinuities unnecessarily across the page. Table 9-1 gives some common examples.

Table 9-1
Typographical Sources of Discontinuities in Scientific Writing

Category	Example	Substitute
abbreviation	e.g.	for example
	fig.	figure
capitals	FORTRAN	Fortran
	APOLLO	Apollo
numerals	19	nineteen
	$13,000,000	$13 million

The first source of discontinuities is needless abbreviation. An abbreviation often includes a period, which is the most powerful piece of punctuation at your disposal. For the reader, a period is a stop sign, a full one-space pause to absorb the idea of the last sentence. Reading a page that is cluttered with needless periods is similar to driving in the heart of a large city—stop, then go, stop, then go. Many of these abbreviations arise from Latin expressions: *exempli gratia* (e.g.), *et alii* (et al.), *et cetera* (etc.), and *id est* (i.e.). Just by knowing their translations ("for example," "and others," "and other things," and "that is"), you can write around them. Some abbreviations such as "in." or "fig." arise from editors who incorporate those abbreviations into their formats. I disagree with this practice, especially for short words. A period is much too valuable to spend on two or three letters.

A second source of typographical discontinuities is using needless capital letters. Using capital letters as abbreviations to stand for a company such as IBM or a complex term such as DNA is often efficient for both the writer and the reader. However, using all capital letters for the names of projects or software is needlessly complex. In the Bible, God settled for initial capital letters. Why should some piece of software deserve better treatment? No single typographical change slows the writing more than placing the letters in all capitals. Why? The reason is that people don't read every letter of a word; rather, people recognize a word not only by the letters it contains, but also by its shape. People recognize a word by seeing the shapes of the ascending letters (*b*, *d*, *f*, *h*, *k*, *l*, and *t*) and the descending letters (*g*, *j*, *p*, *q*, and *y*). However, when words are presented in all capitals, the distinguishing shapes from ascenders and descenders are lost.

A third source of discontinuities comes from needless numerals. Numerals are actual figures: 0, –1, 2.76, 3000. Because numerals make the writing appear more complex than words do, you should use numerals only when necessary. For instance, when numbers can be expressed in one or two words, write them out.

one	two thousand
thirteen	seventy-six

There are a number of exceptions to this rule. Some have arisen to make formats clear—for example, page numbers (page 21) and figure numbers (Figure 2). Others have arisen to make the mathematics clear—for example, negative numbers (–1) and decimals (0.3). In scientific writing, some other important exceptions to this rule are as follows:

specific measurements	12 meters/second
percentages	15 percent
monetary figures	$3,450
large numerals	46 million

Another convention that occurs in English is to avoid beginning a sentence with a numeral. An improper beginning then would be

> 64.1 milligrams of copper corroded during the tests.

If a numeral is called for, then restructure the sentence so that the numeral doesn't appear first:

> During the tests, 64.1 milligrams of copper corroded.

Although these typographical changes seem small, their collective effect is large. Consider an example from a writer who was insensitive to these changes:

> Long Valley, near Mammoth Lakes, CA, is a caldera that was formed c. 990,000 yrs. ago.

Ridding the sentence of needless abbreviation, capital letters, and numerals gives a much more fluid sentence:

> Long Valley, near Mammoth Lakes, California, is a caldera that was formed nearly 1 million years ago.

Incorporating Equations. Often in scientific writing, the most efficient way to convey relationships is through a mathematical or chemical equation. Without equations, even simple relationships can appear needlessly complex:

> The absorptance is calculated as one minus the correction factor times the measured reflectance.

This engineer has made his readers work too hard. Standing alone, this sentence is a puzzle, but in the middle of a document, this sentence is a wall preventing the audience from understanding the work. An equation would have communicated the relationship much more efficiently:

> The absorptance (A) is calculated by

$$A = 1 - kR,$$

where k is the correction factor and R is the measured reflectance.

Although equations simplify relationships, they still make for difficult reading. Readers must stop and work through the meaning of each variable. Therefore, anytime you introduce equations into your writing, you should show why those equations are important. In other words, you should give readers incentives to push through the work of understanding your equations.

The reaction $O_2 + H \rightarrow O + OH$ is the single most important chemical reaction in combustion; it is responsible for the chain-branching of all flame oxidation processes.

Because equations are difficult to read, you want to make them as clear as possible. One thing you should do is define all the terms in your equations. Don't assume that all your readers will read π as the gas phase reaction order or ∂ as the ratio of unburned to burned temperatures:

The burning rate (Ω) of a homogeneous solid propellant is given by

$$\Omega = \frac{\rho}{\alpha}(2\lambda\tau)^{1/2} Le^{n/2}\left[\frac{c(1-\sigma)}{(c-\sigma)(1+\gamma_x)}\right],$$

where ρ, c, and λ are the gas-to-solid ratios for density, heat capacity, and thermal diffusion; τ is a scaled ratio of the rate coefficient for the gas phase reaction (reaction order n) to that of surface pyrolysis (reaction order 0); γ_s is the ratio of the heat of sublimation to the overall heat of reaction; Le is the gas phase Lewis number; α is the fraction of the propellant that sublimes and burns in the gas phase; and σ is the thermal capacity [Margolis and Armstrong, 1985].

Besides stating the importance of equations and defining all variables, you should also clearly state the assumptions of your equations. Readers have to know when your equations apply and when they don't:

> The Townsend criterion for the electrical breakdown of a gas is given by
>
> $$\frac{\omega}{\alpha}\left(e^{\alpha d} - 1\right) = 1$$
>
> where α is the primary Townsend ionization coefficient, ω is the secondary Townsend ionization coefficient, and d is the distance between the electrodes. The criterion for breakdown is actually a physical interpretation of conditions in the gas rather than a mathematical equation because Townsend derived his theory for a steady state case that had no provisions for transient events such as breakdown.

Without the caution in the last sentence, someone could misunderstand Townsend's criterion.

When presenting equations, you should also consider using examples to give readers a feel for the numbers involved:

> In the first stage (Stage I) of an electron avalanche, diffusion processes determine the avalanche's radial dimensions. The avalanche radius r_d is given by
>
> $$r_d = (6Dt)^{1/2}$$
>
> where D is the electron diffusion coefficient and t is time. In the regime of interest (where voltages are greater than 20 percent of self-breakdown), the time of development is so short that little expansion occurs. For the case of nitrogen at atmospheric pressure, $D \approx 860\ cm^2/sec$, $t \approx 10^{-8}$ sec, and $r_d \approx 7.2 \times 10^{-3}$ cm. [Kunhardt and Byszewski, 1980]

Although you might have shown an equation's importance, identified its symbols and assumptions, and given examples, you still may not have done enough. For clear writing, you should convey the physical meaning of the equation. As the physicist Paul Dirac said, "I understand what an equation means if I have a way of figuring out the characteristics of its solution without actually solving it." That is the type of understanding you want to give your readers:

In 1924, Louis De Broglie made an astute hypothesis. He proposed that because radiation sometimes acted as particles, matter should sometimes act as waves. De Broglie did not base his hypothesis on experimental evidence; instead, he relied on intuition. He believed that the universe is symmetric. Using Einstein's energy relation as well as the relativistic equation relating energy and momentum, De Broglie derived an equation for the wavelength of matter, λ_m:

$$\lambda_m = h/p_m.$$

In this equation, h is Planck's constant and p_m is the matter's momentum. In 1927, Davisson and Germer experimentally verified De Broglie's hypothesis.

This discussion could have introduced De Broglie's equation with a detailed mathematical derivation. Instead, the discussion presented De Broglie's argument for symmetry. A rigorous mathematical derivation is not always the best way to communicate an equation.

In derivations of equations, discontinuities often arise from weak transitions between steps. Although you often have to compress derivations to fit the format, you should not take needless jumps:

Our equation for intensity I is

$$I = \int_{-1}^{+1} \frac{dx}{\sqrt{1 - x^2}(1 + x^2)} .$$

Using elementary techniques, we then arrive at

$$I = \frac{\pi}{\sqrt{2}} .$$

This physicist not only lost the continuity in the derivation, but also angered many readers who didn't immediately recognize how to integrate the equation. For almost the same number of words, the physicist could have written

Our equation for intensity *I* is

$$I = \int_{-1}^{+1} \frac{dx}{\sqrt{1 - x^2}(1 + x^2)} .$$

Using a contour integral, we then arrive at

$$I = \frac{\pi}{\sqrt{2}} .$$

Stating how the equation was solved cost the physicist nothing, but saved his readers work.

References

Eichelberger, J. C., "Modeling Mount St. Helens Volcanic Eruption," *Sandia Technology*, vol. 7, no. 2 (June 1983), p. 3.

Kunhardt, E. E., and W. W. Byszewski, "Development of Overvoltage Breakdown at High Pressure," *Physical Review A*, vol. 21, no. 6 (1980), pp. 2069–72.

Margolis, S. B., and R. C. Armstrong, "Two Asymptotic Models for Solid Propellant Combustion," *Sandia Combustion Research Program Annual Report* (Livermore, CA: Sandia National Laboratories, 1985), chap. 5, pp. 6–8.

Saarinen, T. F., and J. L. Sell, *Warning and Response to the Mount St. Helens Eruption* (Albany: State University of New York, 1987).

Chapter 10

Illustration: Making the Right Choices

The most beautiful thing we can experience is the mysterious. It is the source of all true art and science.

—Albert Einstein

There are two types of illustrations: tables and figures. Tables are the arrangements of numbers and descriptions in rows and columns. Figures are everything else: photographs, drawings, diagrams, graphs. This chapter discusses the advantages and disadvantages of each type. When presenting numerical data, you should assess the advantages of tables and graphs before making a choice. When presenting an image, you should assess the advantages and disadvantages of photographs, drawings, and diagrams.

Choosing Tables

Tables have two important uses. First, tables can be used to present numerical data. There are advantages to presenting numerical data in a table, as opposed to using a graph. First, with a table, you can include a high degree of accuracy for the data. For example, with a table such as Table 10-1, you can present data with seven significant digits, an accuracy unobtainable with a line graph. Another advantage of a table is that because of the white space, readers have an easier time locating specific reference numbers, such as the rest mass of a neutron.

Table 10-1
Measured Constants

Name of Constant	Symbol	Measured Value
Avogadro's number	N_0	6.022117×10^{23}/mole
Elementary charge	e	1.602192×10^{-19} coulombs
Planck's constant	h	6.626196×10^{-34} joules-sec
Rest mass (electron)	m_e	9.109556×10^{-31} kilograms
Rest mass (proton)	m_p	1.672614×10^{-27} kilograms
Rest mass (neutron)	m_n	1.674920×10^{-27} kilograms
Speed of Light	c	2.997925×10^{8} meters/sec

A second important use for tables is to present short parallel descriptions that otherwise would have to be listed in the text. Table 10-2 presents the sequence of events that led to the accident at the Chernobyl Nuclear Power Plant [Wolfson, 1991]. Notice that if an author tried to present this information only with text, he or she would end up presenting parallel information in repetitive sentence patterns that would quickly tire the readers. Not only would the repetitive sentence and paragraph patterns take the audience longer to read, but the relationships between the power levels and the events would not be as clear.

Table 10-2
Sequence of Events in the Chernobyl Accident

Date	Time	Power (MW)	Event
4/25	1:00 am	3200	Operators begin power descent
4/25	2:00 pm	1600	Power descent delayed 9 hours Emergency core-cooling system disconnected
4/25	11:10 pm	1600	Automatic control switched off Power descent resumed
4/26	1:00 am	30	Power minimum reached
4/26	1:19 am	200	Operators pull rods beyond limits Two more coolant pumps started Coolant flow limits violated
4/26	1:23 am	2,000,000	Power surges by a factor of 10,000 in 5 seconds

Choosing Figures

Figures include graphs, photographs, drawings, and diagrams. Figures have two important uses. First, with graphs, you have an imagistic way to present numerical data. Second, with photographs, drawings, and diagrams, you have three different ways to present an image.

Graphs. Graphs are drawings that show general relationships in data. Line graphs are the most common type of graph in scientific writing. Figure 10-1 presents the heat transfer from a silicon mask that is cooled with helium gas. Notice that the degree of accuracy in this graph is at best two significant digits. A table could present more.

However, because the engineer wanted to show the general agreement of the experimental data with the theoretical curve, a graph was more effective.

In Figure 10-1, the experimental values are plotted as points, and the analytical solution is represented by a curve. In scientific writing, there are many such conventions associated with line graphs. Some conventions arise for reasons of clarity. For instance, on the axes of a line graph, convention requires that you label the units and designate the scale. Some conventions, such as plotting the independent variable on the x-axis, arise for historical reasons. Other conventions, such as plotting the data on a logarithmic scale, depend on the situation. How much do the data vary over a wide scale? How important is it for readers to see the slope of the curve? For a detailed discussion of the conventions associated with line

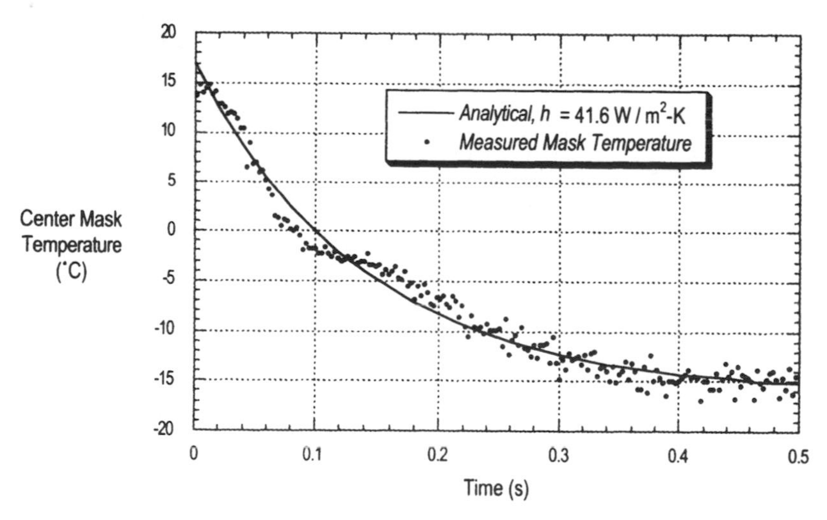

Figure 10-1. Temperature at the center of a silicon mask as a function of time [Laudon and others, 1995]. The mask was first heated by a plasma and then cooled with helium gas. The variable h is the heat transfer coefficient.

graphs, see *The Visual Display of Quantitative Data* by Edward R. Tufte [1983].

Another common graph in science and engineering is the contour plot. Figure 10-2 presents a contour plot of the steady-state temperature across a silicon mask during the plasma heating. Contour plots are excellent for giving a physical feel for how variables such as temperature vary over a surface.

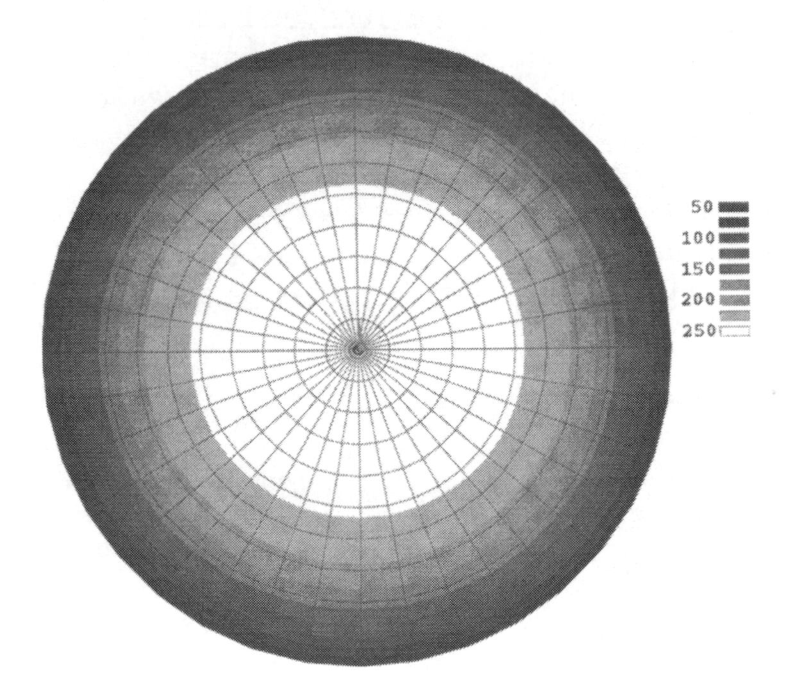

Figure 10-2. Steady-state temperatures (°C) across a silicon mask that is heated by a plasma [Laudon and others, 1995]. Scientists use this mask in x-ray lithography to pattern circuits on silicon wafers.

Other kinds of graphs include bar graphs and pie graphs. Bar graphs compare sizes of different elements. Figure 10-3, for example, compares the natural sources of radiation that the typical person living in the United States experiences. While bar graphs allow you to compare different whole quantities, pie graphs allow you to compare parts of a single whole. Figure 10-4, for example, breaks down the contribution of radiation from all sources (natural and man-made) on the typical person in the United States. An advantage of bar graphs and pie graphs is how well they show dramatic differences, such as the contribution of radiation from natural sources compared to that from nuclear power plants.

Like tables, bar graphs and pie graphs can show numbers to several significant digits of accuracy. You should not assume, however, that bar graphs and pie graphs can replace tables for showing comparisons. When comparing many elements—for example, the mass absorption coefficients of the periodic elements—bar graphs and pie graphs become too busy.

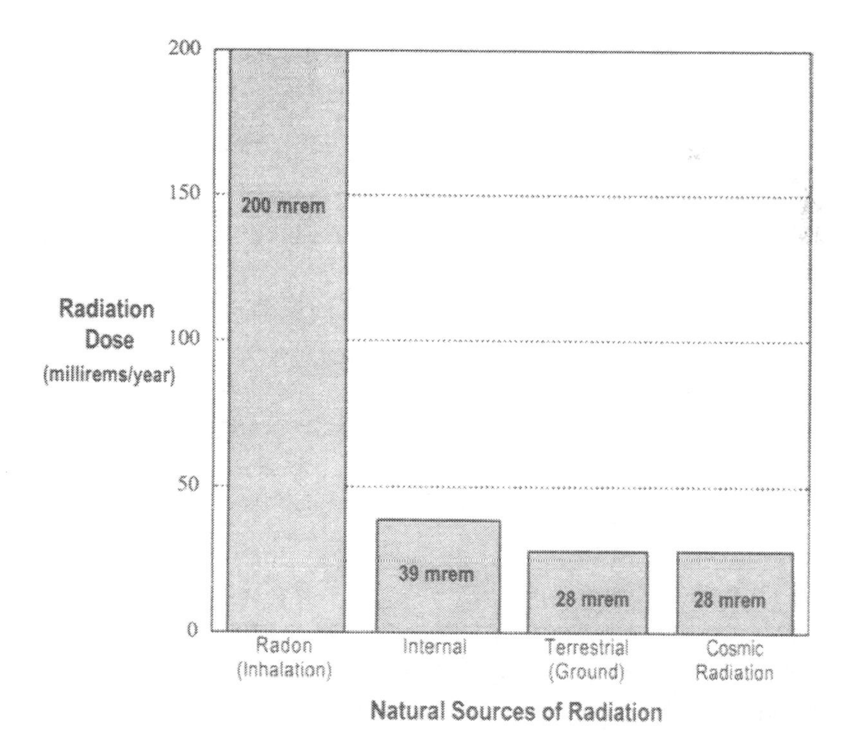

Figure 10-3. Sources of natural radiation that the average person in the United States receives each year [*Radiation*, 1994].

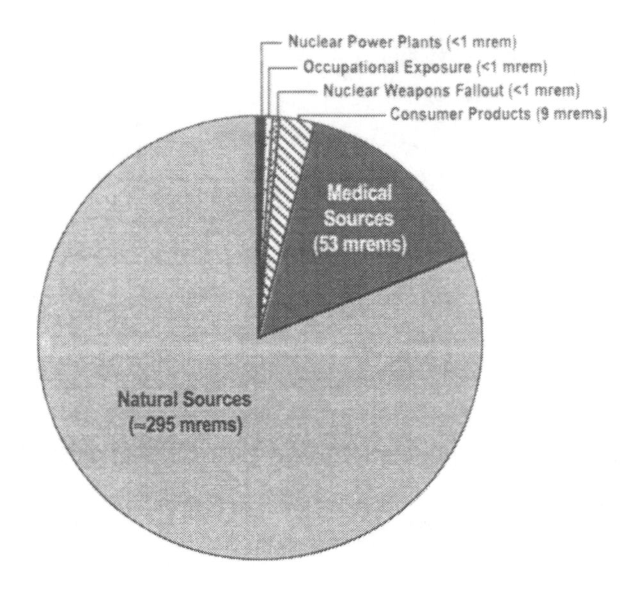

Figure 10-4. The breakdown of annual radiation dosage to the average person in the United States from all sources [*Radiation*, 1994]. The total is about 360 mrem per year.

Photographs. Photographs give readers realistic depictions of images and events. The major advantage of photographs is realism. For instance, once readers see a photograph of the eruption of Mount St. Helens (Figure 10-5), they know that the volcano has, in fact, erupted. For that reason, photographs are important for showing events that have occurred and designs that have been constructed.

This advantage of realism is also a disadvantage. A photograph not only shows a subject's true textures and tones, but also its true shadows and scratches. These extraneous details can easily confuse readers. Consider Figure 10-6, a photograph of a nuclear fusion experiment. Although the photograph is impressive and shows that this fusion experiment actually exists, there are many details such as the cranes and wiring that make it difficult to see how the electrical energy is converted to fusion energy.

Figure 10-5. Eruption of Mount St. Helens in 1980 (courtesy of the United States Geological Survey).

Drawings. Drawings include line sketches and artists' renditions. The major advantage of drawings is that you can control the amount of precision. In drawings, unlike photographs, if there's a detail that distracts from the discussion, you can easily delete it. Figure 10-7, for example, is a depiction of the nuclear fusion experiment shown in Figure 10-6. With photography, you are saddled with having to show the cranes and wiring. However, in a drawing, those details are easily deleted. Also, notice that the drawing presents an unusual perspective (a cutaway) that captures the inner detail of this experiment—something

Figure 10-6. Nuclear fusion experiment at Sandia National Laboratories. Here, an accelerator focuses lithium ions onto deuterium-tritium pellets to produce nuclear fusion [VanDevender, 1985].

a photograph cannot do. Don't assume, however, that the photograph is without value. As stated earlier, a photograph of a test facility makes it clear to the reader that the facility does, in fact, exist.

Besides allowing you unusual perspectives such as cutaway, exploded, or inset, drawings permit you to capture events that cannot be photographed. Figure 10-8, for example, shows a futuristic perspective of a satellite shooting a temperature probe into a passing comet. This drawing appeared in a proposal that requested funds to perform this experiment.

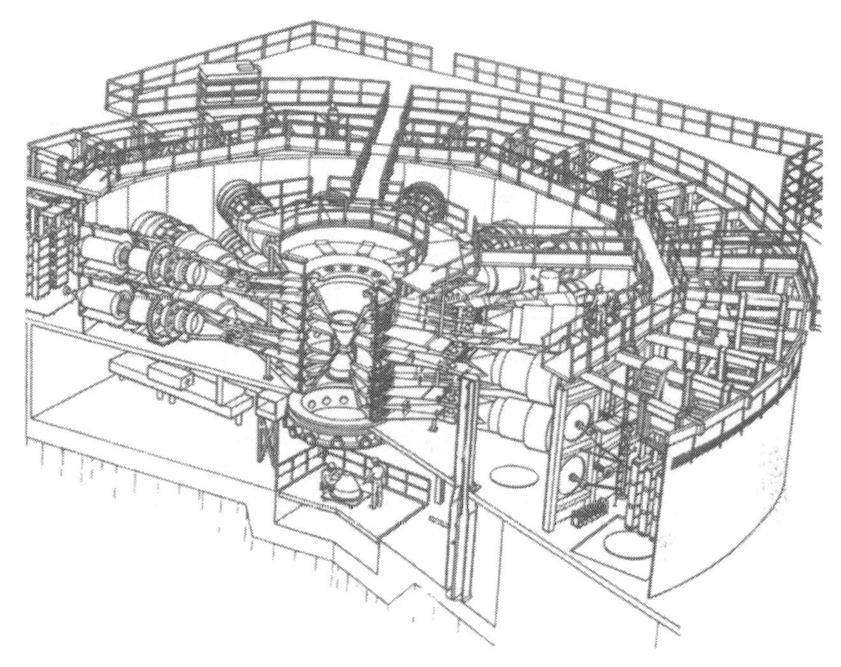

Figure 10-7. Cutaway view of a nuclear fusion experiment at Sandia National Laboratories [VanDevender, 1985].

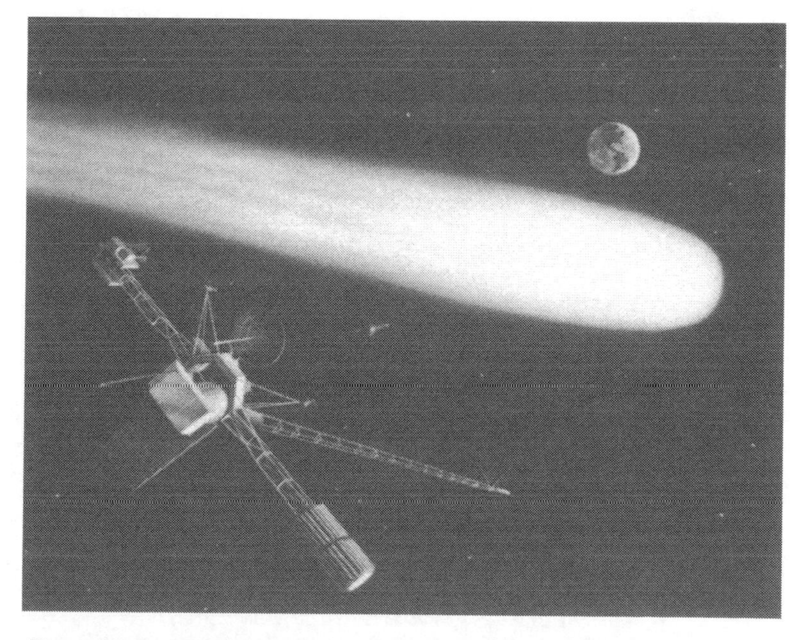

Figure 10-8. Futuristic drawing of a satellite shooting a temperature probe into a passing comet [Young, 1985].

Diagrams. Diagrams are drawings, such as electrical schematics, that communicate through symbols and do not try to depict an object's physical characteristics. The principal advantage of diagrams is that they show how the different parts of a system relate to one another. For example, Figure 10-9 shows the energy flow through the nuclear fusion experiment of Figures 10-6 and 10-7. Here, the diagram shows that the fusion process begins with storing electrical energy in Marx generators and ends with particle beams being focused onto deuterium-tritium pellets. When using a diagram, you have to make sure that your readers know what the symbols represent. Otherwise, the details of the diagram float unanchored in the reader's mind.

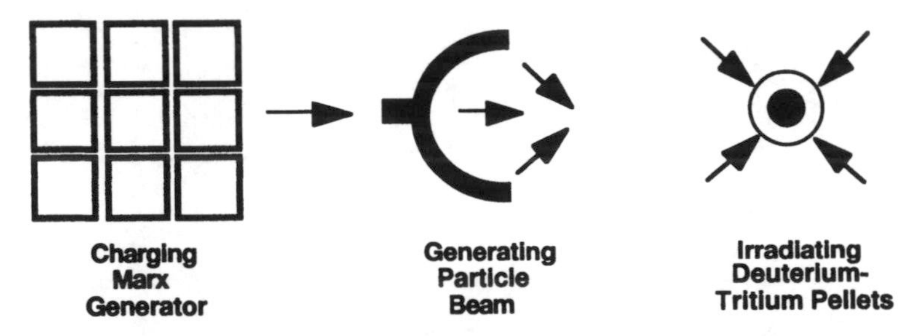

Figure 10-9. Energy flow diagram through the nuclear fusion experiment of Figure 10-6 and Figure 10-7 [VanDevender, 1985].

References

Einstein, Albert, in "Wie ich die Welt sehe," *Mein Weltbild* (Amsterdam: Querido, 1934), pp. 11-17. Original text: *"Das Schönste, was wir erleben können, ist das Geheimnisvolle. Es ist das Grundgefühl, das an der Wiege von wahrer Kunst und Wissenschaft steht."*

Laudon, M. F., K. A. Thole, R. L. Engelstad, D. J. Resnick, K. D. Cummings, and W. J. Dauksher, "Thermal Analysis of X-ray Membrane in a Plasma Environment," presentation, *The 39th International Conference on Electron, Ion and Photon Beam Technology and Nanofabrication* (Phoenix, Arizona: May 1995).

Radiation and Health Effects: A Report on the TMI-2 Accident and Related Health Studies, 2nd edition (Middletown, Pennsylvania: GPU Nuclear Corporation, 1986), chap. 1, p. 1.

Tufte, Edward R., *The Visual Display of Quantitative Data* (Cheshire, Connecticut: Graphics Press, 1983).

VanDevender, P., "Ion-Beam Focusing: A Step Toward Fusion," *Sandia Technology*, vol. 9, no. 4 (December 1985), pp. 2–13.

Wolfson, R., *Nuclear Choices* (Cambridge, Massachusetts: The MIT Press, 1991).

Young, Jack, "Satellite Firing Probe Into Comet Kopff," *SL-21518* (Livermore, California: Sandia National Laboratories, 1985).

Chapter 11

Illustration: Creating the Best Designs

Everything should be as simple as it can be, yet no simpler.
—Albert Einstein

The goals for illustrations are similar to the goals for language. First, you should make your illustrations precise. Precision does not mean having the most accurate illustrations. Rather, precision means having illustrations that best reflect the accuracy of the language. An equally important goal of illustrations is clarity. Too often in scientific writing, illustrations confuse rather than inform. Clarity demands not only that your illustrations convey the information intended, but also that they not convey any unintended information that misleads or confuses the audience.

Another important goal for illustrations is to be fluid. Having an illustration in the middle of your document creates a natural discontinuity for readers. Readers must move their eyes from the text to the illustration. Therefore, you should smooth that transition. Besides striving for precision, clarity, and fluidity, you should anchor your illustrations in the familiar. Being familiar means anchoring an unusual image so that the audience gains a sense of its size and orientation.

Being Precise

Einstein said to keep the information as simple as possible, yet no simpler. This advice applies to choices of precision in illustrations. In your illustrations, avoid details that aren't either self-explanatory or explained in your text. The precision of your illustrations should reflect the precision of your language. A common mistake in scientific writing is to present a figure that is much more complex than the text:

> The thermal storage system stores heat in a huge, steel-walled insulated tank. Steam from the solar receiver passes through heat exchangers to heat a thermal oil, which is pumped into the tank. The tank then provides energy to run a steam generator that produces electricity. Figure 11-1 shows a schematic of this system.

Figure 11-1 is just too complex for the accompanying text. Perhaps you could include this schematic in an appendix for readers who are already familiar with the system, but in the general discussion, you should use an illustration such as Figure 11-2.

Notice that Figure 11-2 includes details that are not explicitly mentioned in the text, namely, the temperatures of the heat transfer fluids. These temperatures, unlike the abbreviations on Figure 11-1, are self-explanatory and serve the document well because the paragraphs following the introduction of this figure discuss these temperatures.

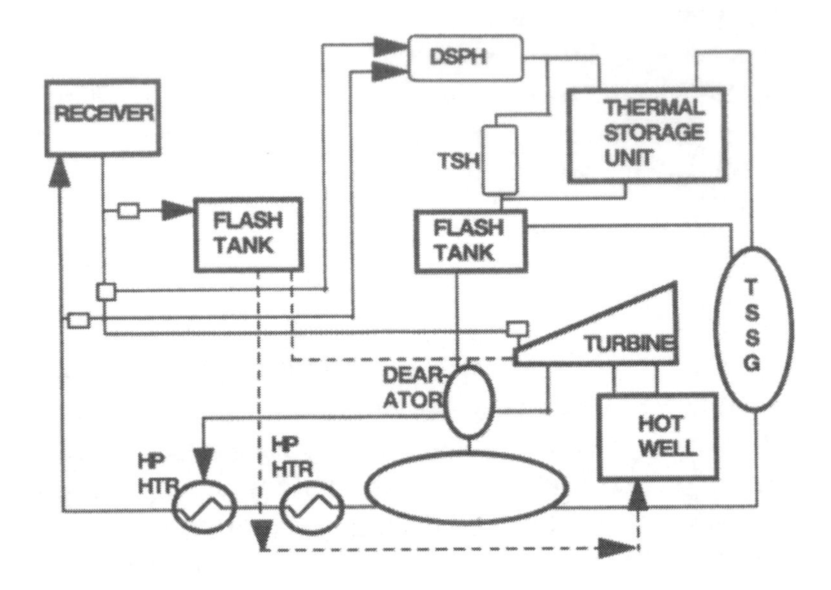

Figure 11-1. Thermal storage system schematic (this schematic is too detailed for the accompanying text).

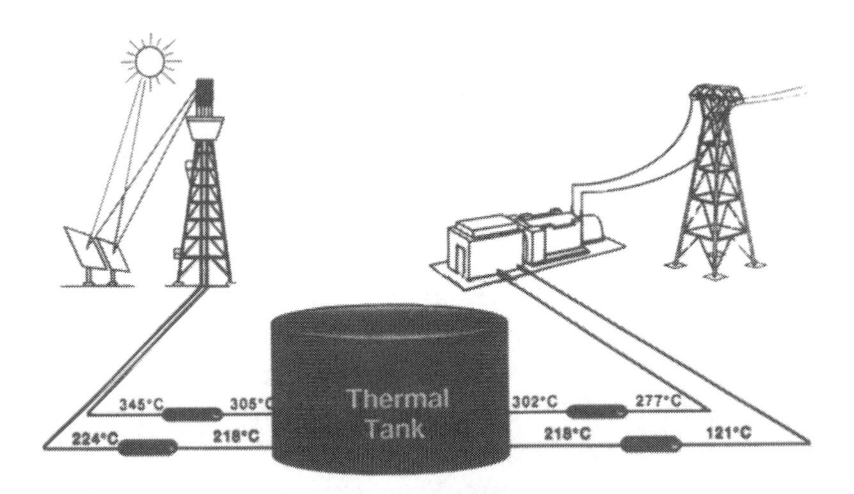

Figure 11-2. Thermal storage system drawing [Radosevich, 1986]. This drawing matches the precision in the accompanying text.

Being Clear

Too often in scientific writing, illustrations confuse rather than inform. Many scientists and engineers mistakenly assume that an illustration is worth a thousand words. That is not so. A picture or photograph may raise more questions than it answers. Figure 11-3 supposedly shows a chemical reaction driven by solar energy, but all you can tell from this photograph is that the reaction is bright. This photograph does not focus attention on the experiment. Instead, extraneous details stand out. For instance, is the lab always this messy? Does this guy really wear that white lab coat? Exactly what is written on the clipboard behind him?

Figure 11-3. Photograph that allegedly shows a chemical reaction driven by solar energy. The background overshadows the central image.

Figure 11-4, on the other hand, is an illustration worth a thousand words. Figure 11-4 shows a parachute system designed for the crew escape module of a fighter jet. Here, the central image of the parachute system stands out with no lab coats or clipboards vying for attention.

Figure 11-4. Parachute system designed for the crew escape module of the F-111 fighter jet [Peterson and Johnson, 1987].

Unclear illustrations arise not only because of mistakes in the selection of illustrations, but also because of mistakes in the language that introduces and labels the illustration. Illustrations cannot stand alone. You need language to mesh the illustration into the document. When you introduce an illustration, you should assign it a formal name and provide enough information, either in the text or in the illustration's caption or heading, so that readers can understand what the illustration is and how it fits into the document.

Besides introducing illustrations in the text, you should identify illustrations with a caption or heading. A well-written caption or heading not only identifies the illustration but also answers immediate questions that the illustration raises. When readers turn a page, their eyes naturally move first toward the illustrations. If those illustrations are interesting, readers often look at the illustrations' captions or titles before continuing to read the text. Therefore, you want your figure captions and table titles to provide enough information to stand independent of the text.

In captioning an illustration, you begin the caption with a title phrase that identifies what the illustration is. In writing this title phrase, you should give the same kind of consideration that you give to the title of the document. In other words, you should choose a title that identifies what the illustration is and provides enough information to separate this illustration from all the other illustrations in the document. Also, in most formats for figures, you have the opportunity to provide an additional sentence or two that clarifies unusual details. In such cases, you should seize that opportunity. The following example shows an introduction and caption for a photograph of a solar power plant. In this example, the writer uses those extra sentences in the caption to explain the unusual bright streaks in the photograph.

> In the Solar One Power Plant, located near Barstow, California, and shown in Figure 11-5, a field of sun-tracking mirrors, or "heliostats," focuses reflected sunlight onto a solar-paneled boiler, or "receiver," mounted on top of a tall tower. Within the receiver, the solar energy heats a transfer fluid that drives a turbine. Electricity is then produced by generators coupled directly to turbines. Also included in this system is a thermal storage system that can operate the plant for several hours after sunset or during cloudy weather. The maximum power output from this plant is 12.2 megawatts.

Figure 11-5. Solar One Power Plant located near Barstow, California [Falcone, 1986]. The bright streaks on either side of the receiver are standby points to prevent the mirror beams from converging in the airspace above the plant when the plant starts up each day.

Being Fluid

For illustrations to be fluid, you have to smooth the transition between what you say and what you show. The most important way to smooth this transition is to match the information in your text with what's in your illustration. You'd be surprised at how often scientists and engineers will say one thing in their text and then present a figure or table that shows something quite different.

> The testing hardware of the missile shown in Figure 11-6 has five main components: camera, digitizer, computer, I/O interface, and mechanical interface. Commands are generated by the computer and then passed through the I/O in-

terface to the mechanical interface where the keyboard of the ICU is operated. The display of the ICU is read with a television camera and then digitized. This information is then manipulated by the computer to direct the next command to the I/O interface.

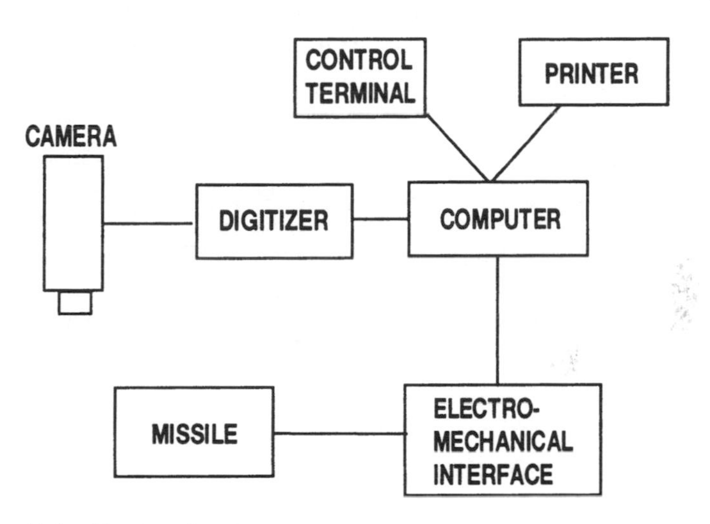

Figure 11-6. Testing hardware of missile (this illustration does not agree with the accompanying text).

This illustration has many weaknesses. For one thing, there are seven parts to the testing system in the figure, not five as stated in the text. Two of the parts (the printer and control terminal) are given names different from what's in the text ("I/O interface"). What's worse, the engineer depicted the missile in the same manner as the testing system so that you can't distinguish the testing system from the thing that's being tested. The missile should be set apart in some way from the five main parts of the testing system. Finally, although this illustration is a diagram, you don't gain a sense of flow through the system. The engineer didn't even aim the camera at the missile.

Consider the following revision with attention to our goals for language and illustration:

Our system for testing the safety devices of the missile consists of four main parts: computer, camera, digitizer, and electromechanical interface to the missile. In this system, shown in Figure 11-7, the computer generates test commands to the missile through the electromechanical interface. The test results are read with a television camera and then digitized. The computer receives the information from the digitizer and then directs the next test command.

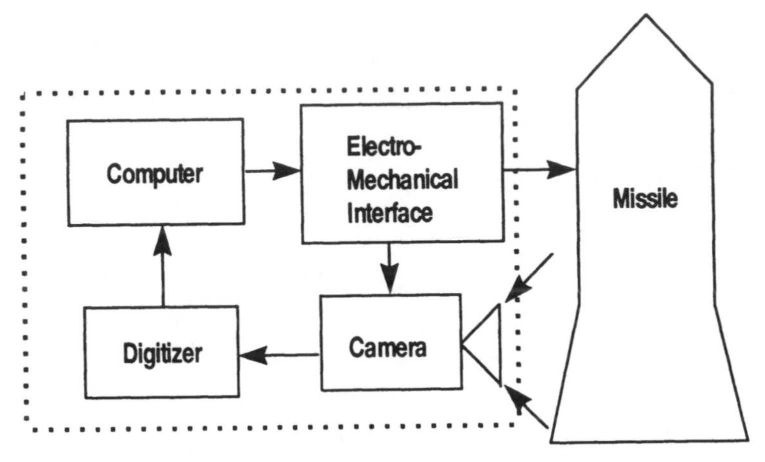

Figure 11-7. System for testing the safety devices of missile (this illustration does agree with the accompanying text).

Being fluid also means placing illustrations so that they closely follow their text references. Many scientists and engineers don't consider their audience when laying out illustrations in the text. Often, these authors just place their illustrations at the end of the document or fit them in where there's some white space. This method may make things easier for the writer, but it causes difficulties for the readers. Readers have a difficult enough time understanding the work without having to wander through the document to find a particular figure or table.

The worst layout mistake that you can make is to place an illustration before its introduction in the text. Readers of scientific documents are not like readers of novels. They often do not read every section in its original order. Even if they do read the sections in order, readers often skim through certain sections. Therefore, misplaced figures and tables can easily confuse readers. An interesting figure that appears before its introduction in the text often causes readers to read backwards, in vain, looking for its introduction. Positioning illustrations before the text introduction of those illustrations is not only confusing writing; it is inconsiderate writing.

Ideally, you'd like to have your illustrations fall just below the paragraph that introduces them. Unfortunately, page breaks often make this arrangement impossible. The next best arrangement then is to have your illustrations follow as closely as possible the paragraph that introduces them. This way, readers can quickly compare the text and illustration, and those readers who start with the illustrations and read backwards won't have far to go.

Finally, don't think, as many scientists and engineers do, that your illustrations have to fill the entire image area of the page. Although you should make illustrations large enough to clarify details, you should also include a reasonable white space border. Remember: White space helps emphasize details.

Being Familiar

Anchoring your illustrations in the familiar is also important in scientific writing. If your text does not answer questions raised by your illustrations, you will frustrate your readers. To make your illustrations familiar, you should consider what your audience does and doesn't know. Because you've spent months, perhaps years, on your project, this task is not easy. Nonetheless, you need to find a way to see your illustrations as your readers see them—with fresh eyes.

Too often, scientists and engineers miss the opportunity to show relative size of an image by including either a scale or familiar-sized image alongside the depicted image. Although an audience may read in the caption of Figure 11-8 that each mirror array has a surface area of 39 square meters, the audience will probably not appreciate that size without the two men in the foreground.

Figure 11-8. Solar mirrors at the Solar One Power Plant [Falcone, 1986]. Each mirror stand has a surface area of 39 square meters.

References

Einstein, Albert, quotation attributed by Hans Byland (1928). Original text: *"Schreibe ich zu kurz, so versteht es überhaupt niemand; schreibe ich zu lang, so wird die Sache unübersichtlich."*

Falcone, P. K., *A Handbook for Solar Central Receiver Design*, SAND86-8006 (Livermore, CA: Sandia National Laboratories, 1986).

Peterson, C. W., and D. W. Johnson, *Advanced Parachute Design*, SAND87-1648C (Albuquerque: Sandia National Laboratories, 1987).

Radosevich, L. G., *Final Report on the Experimental Test and Evaluation Phase of the 10 MWe Solar Thermal Central Receiver Pilot Plant*, SAND85-8015 (Livermore, CA: Sandia National Laboratories, 1986), p. 43.

Chapter 12

Writing Correspondence

I have made this letter longer than usual because I lack the time to make it shorter.

—Blaise Pascal

Correspondence is an effective way to make requests and deliver specific information. When you respond to a job announcement, you write a letter. When you summarize a staff meeting, you write a memo. When you announce a sudden change in a schedule, you send an electronic mail message. Unlike a telephone conversation, correspondence presents the audience with a contract that is dated and can support a claim in court. Besides the advantage of legality, correspondence has the advantage of efficiency when you are trying to reach several people. Writing a single memo will take you less time than phoning twenty-five people. Also, when the message is difficult to phrase, such as a negotiation point in a contract, correspondence allows you the advantage of revising the message until it is correct. Finally, from the audience's perspective, correspondence has the advantage that it can be reread.

The writing of most correspondence is straightforward. For instance, in writing a letter to order equipment, the correspondence simply entails that you state what you want, present the necessary information, and then close. Other correspondence is more difficult. For example, suppose that you are a manager reviewing a paper written by one of your engineers. If the work is promising, but has many weaknesses that must be addressed, you have to walk a thin line in your review. On the one hand, you want to make the engineer aware of the problems and the work that still remains. On the other hand, you don't want to discourage the engineer so much that he or she gives up on a worthwhile project.

Constraints of Correspondence

Before beginning to write a letter or memo, you should consider the writing constraints. With correspondence, two interesting writing constraints to consider are audience and mechanics. As stated earlier, there are four general questions to consider with audience: (1) who is the audience? (2) what does the audience know about the subject? (3) why is the audience reading the document? and (4) how does the audience read the document?

With correspondence, the first question, about the identity of the audience, is easy to establish because you generally write correspondence to one type of audience, and many times to just one person. The second question, about the audience's knowledge of the subject, requires more thought. As with an article or report, the answer to this question dictates how much depth you achieve, how much background information you provide, and which words you define.

The third question, about why the audience reads the correspondence, is often answered in two parts: why the audience begins reading the correspondence and why the audience continues reading the correspondence. Audiences begin reading correspondence for the simple reason that it is in their mail slots. Part of the motivation is obligation to professional duty; another part is curiosity. Why the audience begins to read a letter or memo and why the audience continues to read the correspondence are two different things. In assessing why the audience continues reading the correspondence, you should consider the fourth question about audience, namely how the audience reads the document.

Most people do not read their correspondence in the same way that they read a long novel—sitting in an easy chair and drinking a cup of tea on a long winter night. Instead, most people read correspondence standing next to the wastebasket or, with electronic mail, poised with a finger on the delete button. Because readers have the correspondence in such a precarious position, writers should tell their readers as quickly as possible what the main points of the correspondence are. Otherwise, the readers might lose patience and discard the correspondence on the grounds that it is "junk" mail.

People read correspondence quickly, much more quickly than they read articles or reports. The reading speed for letters and memos corresponds to a sprint, as opposed to a steady jog for articles and reports. For that reason, having multiple pages on correspondence intimidates readers. Although there will be occasions in which the scope of the correspondence requires multiple pages, you should strive for correspondence no longer than one page.

Besides audience, mechanics is another important constraint in writing correspondence. A mechanical error such as a run-on sentence or a misspelling in a one-page memo will receive more attention than the same error in the middle of a fifty-page report. Each mechanical error costs the writer a notch of credibility with the reader. For important correspondence, such as a job application letter, a mechanical error can be even more expensive. Employers who are swamped with applications look for quick ways to reduce the stack, and finding mechanical errors in the application letters is a common means for doing so.

To eliminate mechanical errors, a good idea is to let the letter or memo or electronic message cool awhile before you make the final proof. Better still, for your important correspondence, let it sit overnight and then proof a printed version the first thing in the morning. Not only will you be better able to catch the difficult proofing errors, such as a missing or repeated word, but you'll be in a better position to evaluate the tone of the letter. For instance, an adverb such as "frankly," which may seem okay at five o'clock at the end of a long day, might sound too cold and distant at eight o'clock the next morning.

Style of Correspondence

Four aspects of style warrant special consideration in the style of correspondence: organization, emphasis, clarity, and forthrightness. The organization for correspondence is similar to the organization of reports and articles, which means that you can think of correspondence as having a beginning, middle, and ending. As in reports and articles, the beginning of a correspondence is important because it determines whether the audience will continue reading. For that reason, you want to orient your readers as quickly as possible in the beginning. That orientation means getting to the point in the first paragraph—the first sentence if possible.

The middle in the correspondence also accomplishes the same task that the middle of a report or article does: delivery of the information. In most correspondence, that delivery translates to writing one or more paragraphs that present the information promised in the beginning.

The correspondence ending, which usually encompasses the final sentence or paragraph of the correspondence, can go in several directions. In a correspondence that primarily informs, the ending often presents a summary of the information. In a correspondence that both informs and persuades, the ending often presents a call for action, in which you state what you want the audience to do. Sometimes, that call for action is loud, as in this closing to a memo proposal: "With your approval, I will begin these experiments." Other times, the call is subtle, as in this closing to a job application letter: "Thank you for your consideration."

For an example of a strong organization in correspondence, consider Albert Einstein's letter to President Roosevelt written in 1939. The letter warned of Germany's rapid development of atomic energy in the Second World War. The letter was difficult to organize for several reasons. First, the science was complex, and much of the work was still at the theoretical stage. Second, the audience was non-technical. Third, decisive action was needed, but in 1939 there was no means, such as Los Alamos National Laboratory, for carrying out that action.

> Some recent work by E. Fermi and L. Szilard, which has been communicated to me in manuscript, leads me to expect that the element uranium may be turned into a new and important source of energy in the immediate future. Certain aspects of the situation which has arisen seem to call for watchfulness and, if necessary, quick action on the part of the Administration. I believe that it is my duty to bring to your attention the following facts and recommendations.
>
> In the course of the last four months it has been made probable—through the work of Joliet in France as well as Fermi and Szilard in America—that it may become possible to set up a nuclear chain reaction in a large mass of uranium, by which vast amounts of power and large quantities of new radium-like elements would be generated. Now it appears almost certain that this could be achieved in the immediate future.
>
> This new phenomenon would also lead to the construction of bombs, and it is conceivable—though much less cer-

> tain—that extremely powerful bombs of a new type may thus be constructed. A single bomb of this type, carried by boat and exploded in a port, might very well destroy the whole port together with some of the surrounding territory. However, such bombs might very well prove too heavy for transportation by air.
>
> The United States has only very poor ores of uranium in moderate quantities. There is some good ore in Canada and the former Czechoslovakia, while the most important source of uranium is Belgian Congo.
>
> In view of this situation you may think it desirable to have some permanent contact maintained between the Administration and the group of physicists working on chain reactions in America. One possible way of achieving this might be for you to entrust with this task a person who has your confidence and who could perhaps serve in an unofficial capacity. His task might comprise the following:
>
> a) to approach Government Departments, keep them informed of the further development, and put forward recommendations for Government action, giving particular attention to the problem of securing a supply of uranium ore for the United States;
>
> b) to speed up the experimental work, which is at present being carried within the limits of the budgets of University laboratories, by providing funds, if such funds be required, through his contacts with private persons who are willing to make such contributions for this cause, and perhaps also by obtaining the co-operation of industrial laboratories which have the necessary equipment.
>
> I understand that Germany has actually stopped the sale of uranium from the Czechoslovakian mines which she has taken over. That she should have taken such early action might perhaps be understood on the grounds that the son of the German Under-Secretary of State, von Weizäcker, is attached to the Kaiser-Wilhelm-Institut in Berlin where some of the American work on uranium is now being repeated.

In addition to organization, emphasis is important to consider in correspondence. One way to accent important details in a letter or memo is to place those details in either the first or last sentence. These two sentences border on more white space than any other sentences, and therefore receive more emphasis. In his letter to Roosevelt, Einstein used the first sentence of his letter to introduce the new energy source that Germany was developing. He used the last sentence to show the strides that Germany had taken to develop that source.

In the language of correspondence, you should pay particular attention to being clear and forthright. Clarity is important in correspondence because audiences read correspondence faster than they read other types of documents. For this reason, you should opt for shorter sentences and paragraphs than you would in a report. Short paragraphs and sentences will reduce the chances that your audience will misread your message. You might ask, why not use short sentences and paragraphs in reports and articles as well? The reason is pacing—an audience cannot sprint for more than a page or so.

In correspondence, the goal of being forthright is important because in correspondence, more than other types of documents, tone is difficult to control. For instance, how in a job application letter do you talk about your accomplishments without sounding boastful? Also, in a letter complaining about faulty workmanship, how do you motivate the reader to repair the damage without alienating the reader? The answers are not simple. Often, scientists and engineers lose control of tone by avoiding simple, straightforward wording. When some people sit down to write a letter or memo, they change their entire personality. Instead of using plain English, they use phrases such as "per your request" or "enclosed please find." Because these phrases are not simple and straightforward, they often inject an undesired attitude, usually arrogance, into the writing.

For an example of correspondence that brings together the four discussed elements of style, consider the following memo [Thole, 1991]:

As you know, the past six months we have not been able to make simultaneous Laser Doppler Velocimetry (LDV) and hot-wire measurements of velocity in the wind tunnel because our LDV seed has contaminated the hot-wire. Titanium dioxide, the seed particle that we use, changes the hot-wire voltage by as much as 2 percent over a 40 minute period—a voltage change that represents a 17 percent difference in velocity. This memo presents the results of my search to find a LDV seed particle that does not contaminate the hot-wire.

After unsuccessfully trying other seed particles such as alumina and silicon carbide, I received an idea from Professor Jim Wallace of the University of Maryland. Professor Wallace uses incense to perform smoke visualization studies while simultaneously making measurements with a hot-wire. Even though Professor Wallace needs a dense smoke for the flow visualization, he claims that his hot-wires are not contaminated by the tar particles contained in the incense. Professor Wallace's smoke generator includes a steel wool filter to collect the tar particles. This generator also cools the smoke to prevent temperature fluctuations from affecting the hot-wire measurements. The smoke is cooled by simply sending the smoke through an air-cooled copper coil before injecting it into the wind tunnel.

After this discussion with Professor Wallace, I began testing incense and found that it was an acceptable LDV seed. The steel wool, in conjunction with the air-cooled copper coil, works as an acceptable filter for the tar particles. From this research, I recommend that we use incense as the seed particle for our simultaneous measurements of velocity using the hot-wire and the LDV.

References

Einstein, Albert, letter to President Franklin Roosevelt (2 August 1939).

Pascal, Blaise, *Lettres Provinciales*, XVI.

Thole, K., memo to Professors David Bogard and Michael Crawford (Austin, Texas: University of Texas, 14 April 1991).

Chapter 13

Writing Proposals

I don't mind your thinking slowly, but I do mind your publishing faster than you think.

—Wolfgang Pauli

A proposal presents a strategy for solving a problem. Giving specific advice about writing proposals is difficult because proposals vary so much. Proposals range from thousand-page reports to one-page forms. Some proposals are solicited by clients bent on solving a specific problem; others involve audiences who are unaware that a problem even exists. In addition to differences in format and audience, the writing situations vary as well. Some proposals are individually written; others are collaborative efforts involving large teams of scientists, engineers, managers, and accountants. Finally, the subject matter of proposals varies greatly—from designing bridges to exploiting genetic weaknesses in a virus.

Besides being difficult to discuss, proposals are also difficult to write. Why? In most science and engineering documents, you write about something that has happened. In these situations, you have results from an experiment or a computational analysis, and those results serve as the foundation for the writing. In a proposal, however, you must imagine a solution to a problem and then write about that imaginary solution. In other words, you write about what you anticipate, rather than what you know. That writing at the "imagined" level, as opposed to the "concrete" level, makes proposal writing difficult.

Another reason that writing proposals is difficult has to do with the primary purpose of proposals. While the primary purpose of reports and articles is to inform, the primary purpose of proposals is to persuade. Your goal in proposal writing is not just to inform your audience about a solution to a problem, but to convince your audience to give you funds so that you can solve a problem.

Constraints of Proposals

Proposals can take days, weeks, sometimes months to put together. On large contracts, it is not uncommon for companies and laboratories to spend tens of thousands of dollars in the preparation of proposals. On top of that, companies and laboratories not receiving the bid must absorb the monetary losses from that preparation. Therefore, you should carefully consider the constraints of proposals before committing yourself and your resources to the writing. The three most important constraints in proposal writing are format, politics, and audience.

Format of Proposals. Although the formats of most unsolicited proposals are left to your discretion, the formats of most solicited proposals are defined in great detail. These definitions include not only the layout and typography, but also the length, the order of information, and the names of the headings and subheadings. More important, failure to comply with the format usually results in the proposal being excluded from consideration.

Why are the formats of solicited proposals so specifically defined? The main reason is that the client wants some guidelines to evaluate all incoming proposals. Often the client has an evaluation form for assessing the proposals, and the order of items on this form corresponds to the order of information requested for the proposal. A more subtle reason that the client specifies such detail in the format is that the client wants to see whether the submitters can follow directions. If the submitters can't follow simple directions on the names of subheading titles, then how can the client expect the submitters to solve the complex problem that generated the proposal request? Similarly, if the submitters can't hold the line on page-length restrictions, then how can the client expect the submitters to hold the line on budget restrictions?

Politics of Proposals. In proposal writing, there are variables outside the quality of the proposal ideas and the quality with which those ideas are presented that determine whether the proposal is accepted. These variables constitute the political constraints on proposals. What kinds of projects is Washington funding this year? Is collaboration between companies, laboratories, or universities an advantage or disadvantage? How important are issues such as experience or geographical region? Although the constraint of politics varies greatly from contract to contract, there is one important point to understand: Many times, situations will arise in which no matter how strong your ideas are and no matter how well you write the proposal, you will not win the contract. In some solicited proposals, for example, the funding agency has someone in mind for the contract from the outset and has issued the proposal request only to fulfill a government regulation. Likewise, in some unsolicited proposals within a company, there might be a policy change in the works that would preclude your proposed idea.

Given that some proposals will fail because of politics, you should find out as much as you can about a proposal situation before you commit your time and resources to the task: Carefully read the request for proposal, call the contact person if you have questions, and honestly evaluate yourself and your ideas against the likely competition. These steps should not inhibit you from writing proposals; rather, they should guide you in deciding which proposal contracts to seek.

Also, given that some proposals will fail because of politics, you should look at the success of your proposals not in terms of the proposals that you write in a single year, but in terms of the proposals that you write over several years. In a pool of several proposals, you will have some that are decided on politics and some that are decided on the intrinsic merit of the ideas and therefore on the quality with which those ideas are presented.

Fortunately, the effort spent on writing one proposal often carries over to the writing of other proposals. For instance, a "Statement of Problem" or "Facility Resources" section that you write for one proposal can often be adapted for use in other proposals. This adaptation does not mean that you simply cut and paste the old text into the new document; instead, it means that you use the old text to help draft the new proposal. You still have to rethink, revise, and polish that old text so that it fulfills the new constraints of the document and meshes with the surrounding sections.

Audiences of Proposals. Besides format and politics, a third constraint that has a special role in proposals is the audience. Unlike the typical audience for correspondence, which consists of one reader or one type of reader, the typical audience for proposals contains at least two types of readers: a management audience and a technical audience. The management audience reviews proposals to see whether the proposed plan is feasible from the perspectives of time and money, while the technical audience reviews proposals to see whether the plan is feasible from the perspectives of science and engineering. Even in a proposal written to one reader, say a manager within a company, that one reader will look at the proposal from both a management perspective and a technical perspective.

In writing a proposal, you have to address both audiences. Perhaps the best way to envision your readers is to put yourself in their positions. For solicited proposals, the request for proposal allows you to do just that. From a well-written request for proposal, you know who the readers are, why the readers have requested the proposal, the level of detail that you should present in your proposal, and how the audience will evaluate the proposal. In other words, you know the answers to the same four questions about audience that you ask yourself before writing a report or article.

As an exercise for understanding the audience for solicited proposals, assume that you are on the other side of the table in the proposal process. In other words, assume that you are the audience, rather than the writer, for a proposal. As an example, imagine that you are a fish-and-wildlife commissioner who oversees a park called Fire Lake. Fire Lake is a fisherman's haven with bass, bream, and trout. Over the past couple of years, though, your commission has received numerous complaints from fishermen stating that the number of fish in Fire Lake has declined. To determine whether these complaints have a basis, your commission has decided to hire someone to count the fish in Fire Lake. Because you don't have a "fish-counter" on contract, you are obligated by state law to accept proposed bids from "fish-counters" across the state. What kind of information do you put in the request for proposal?

Well, one thing that you would want to know is the cost. You wouldn't want to award the contract to someone who has a blank check with your signature on the bottom. If you did and if you hired someone who was unscrupulous, you might receive a bill that severely dents your budget. Besides cost, you would want to know the schedule for the work so that you are sure that the work will be done in a timely manner. Otherwise, you might end up with a "hippie" fish-counter who comes in with a "low-ball" bid of only sixty dollars a week until the job is completed. The problem with the hippie fish-counter is that the job never gets completed. The hippie fish-counter is content to live on sixty dollars a week. In fact, he plans to pitch a tent on the banks of old Fire Lake, listen to the Grateful Dead, and count fish—indefinitely.

A third aspect for the proposal request would be the type of information that the fish-counter actually presents. Will the fish counter present just the total number of fish in the lake, or the number of fish within each species? The number of fish in each species is much more useful. For one thing, you could have about the same number of fish as you did twenty years ago, but if all those fish are carp rather than an assortment of bass, bream, and trout, then you've got a problem. Besides knowing the number of fish within each species, you would want to know how accurate the numbers are. Within 10 percent? Within 20 percent? Along these same lines, you would want to establish guidelines for an acceptable accuracy.

There was a case in San Francisco in which the city hired a consultant to estimate how many people would show up for the 50th anniversary of the Golden Gate Bridge. For the celebration, the city had decided to close the bridge to automobile traffic and allow people to walk across, just as they had done during the grand opening fifty years before. The consultant who won the contract did some surveys and announced a few weeks before the celebration that about 50,000 people would show up to walk across the bridge [Nolte, 1987]. On the basis of that number, the city ordered an appropriate number of buses and policemen. In addition, the city allotted four hours for the bridge to be closed to automobile traffic and had orange cones placed in the center of the bridge, so that pedestrians from the San Francisco side of the bridge could walk on one side of the cones and pedestrians from the Marin side could walk on the other.

The morning of the celebration came, and instead of 50,000 people arriving, over 800,000 people arrived. Because there weren't enough buses to handle all the people wanting to get to the bridge, the streets near the bridge became jammed with double and triple parking. Worse yet, the bridge became gridlocked with people and could not be cleared. In fact, the bridge became so crowded that people started climbing onto the scaffolding of the bridge because their claustrophobia overcame their acrophobia. Afterwards, the city petitioned to get back the money from the consultant who had given the inaccurate crowd estimate, but because the proposal request had no specifics about accuracy, the courts ruled that the consultant could keep the money.

A fourth consideration for your proposal request would be finding out the methods that the fish-counter will use. Otherwise, you might hire a psychopathic fish counter who presents you with a low bid ($200), a reasonable schedule (2 days), and an incredible accuracy (within one fish). What you wouldn't realize without knowing the methods is that this psychopathic fish counter plans to purchase $100 worth of poison, dump it in Fire Lake, wait a couple of days, and then announce that you have exactly zero fish.

Style of Proposals

An unusual stylistic aspect of proposals is their overall organization. Unlike other documents, in which a beginning-middle-ending organization exists, the organization for proposals consists essentially of two parts. The first part presents a hole that needs to be filled. This hole is a statement of the problem that has produced the need for a proposal. The second part of the organization is the piece that fits into the hole. This part is the proposed plan for addressing the problem. In its actual form, the proposal will often have the same section names as other documents: title, summary, conclusion, and so forth. However, it is the relationship of the original problem to the proposed solution that establishes the organization for the document.

Problem Statement. The first part of the proposal's organization, often titled the "Statement of Problem," shows that the writer understands the problem that has generated the proposal. This section is taken for granted by many scientists and engineers, although it is highly valued in the evaluation of proposals, especially unsolicited proposals and research proposals to funding agencies such as the National Science Foundation. In these latter situations, this section not only shows that the writer understands the problem, but also makes the audience aware of how important the problem is.

In stating the problem, you should show, not just tell, the audience that there is a problem. For example, if you desire funding to study ways to reduce the space debris that is orbiting the earth, then you would have to show first that there is a substantial amount of debris in orbit and second that this debris poses a problem. In the following excerpt [Wright, 1991] from such a proposal, notice how the author uses example after example to back up his points.

> Currently, NASA is tracking over 7000 artificial satellites in orbit about the earth [Baker, 1990]. Of these satellites, shown in Figure 13-1, only 6 percent represent functioning satellites. The other objects orbiting the earth are considered space debris. The smallest size being tracked is 10 centimeters in diameter, or about the size of a baseball [Subcommittee, 1989]. In addition to the debris being tracked, there exist an additional 30,000 to 70,000 objects between 1 to 10 centimeters that are too small for tracking [Baker, 1989].
>
> The debris includes operational debris, which is produced during normal space activity. Sources of operational debris include experimental equipment, astronaut possessions, and human refuse [Johnson and McKnight, 1987]. For example, in the early 1960s, an experiment called the "Westford needles experiment" released numerous copper

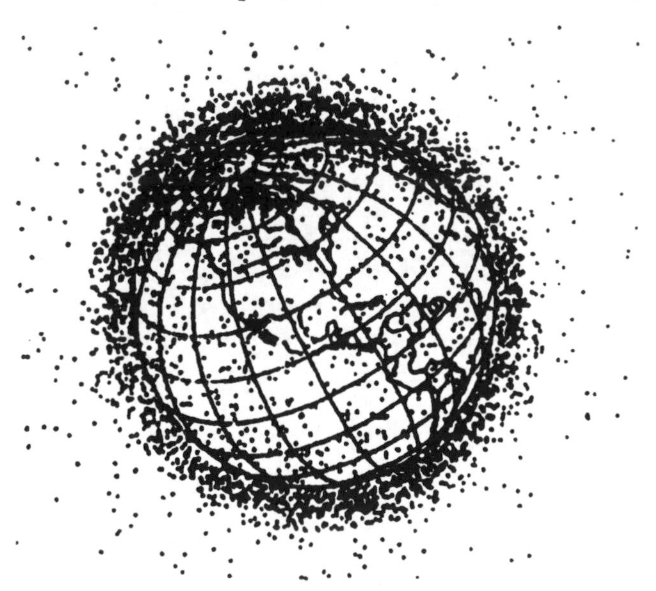

Figure 13-1. Computer representation of the more than 7000 artificial satellites that are being tracked in their orbits about the Earth (representation courtesy of Teledyne Brown Engineering). Less than 6 percent of these satellites represent functioning satellites; the rest are space debris.

needles in orbit about the Earth. The experiment's goal was to release large quantities of small copper dipoles resonant at x-band and radio frequencies at a moderate altitude of 3,900 kilometers. Not only did the experiment not work, but many of those needles are still in orbit today. Operational space debris has also been created by astronauts. For instance, during the first American space walk in June 1965, one of Ed White's gloves drifted out of the Gemini spacecraft. In another mission, a Challenger astronaut lost a powered screwdriver and several screws while repairing the Solar Maximum satellite [DeMeis, 1987]. In still other situations, Gemini and Soviet astronauts threw bags of garbage overboard [Baker, 1989].

Besides operational debris, the orbital space debris also includes microparticulate matter, such as solid-propellant fuel particles, paint flakes, and thermal coating particles. Compared with the large objects of operational debris, the microparticulate matter might appear to be a negligible threat. However, these particules can obtain high orbital speeds in space that increase their threat to space activity. For instance, a paint chip caused a 4 millimeter crater in a window during a 1983 mission of the Space Shuttle Challenger [DeMeis, 1987]. This particle was travelling at 3 to 6 kilometers per second and because of the severity of the impact, the window could not be reused [Baker, 1989]. In another example, one impact from a particle penetrated three layers of a seventeen-layer blanket retrieved from the Solar Maximum satellite. Analysis of the impact crater showed that the particle was probably an ice crystal from the Shuttle waste management system. Scientists fear that these particles could cause much greater damage. For example, a paint chip travelling at 10 kilometers per second could puncture an extravehicular space suit [DeMeis, 1987].

This problem statement was straightforward to develop because the writer required only two steps to make a case for removing orbital debris: (1) showing that a lot of orbital debris exists, and (2) showing that this orbital debris was dangerous to future space missions. Consider another example in which the proposed work is not quite so easy to justify. In this example, notice how the writers [Thole and others, 1991] methodically move from one point to another until a bridge is formed between a widely recognized problem (increasing the efficiency of airplane engines) to one not nearly so recognized (finding a way to produce high turbulence).

> In gas turbine engines, such as jet airplane engines, the higher the combustion temperature, the higher the engine efficiency. At present, the combustion temperatures desired for these engines are much higher than the melting temperatures of the blades in the engines. For example, a common situation is a turbine blade with a melting temperature of 1000K operating in conditions in which the turbine inlet temperatures are 1700K [Byworth, 1986]. To achieve these higher combustion temperatures, the turbine blades are cooled.
>
> A common method for cooling the turbine blades is film cooling. In film cooling, a cold fluid is injected through the blade surface. This injection results in a cooling film between the hot combustion gases and the metal surface of the blade. The typical geometry of a film-cooled turbine blade is shown in Figure 13-2.

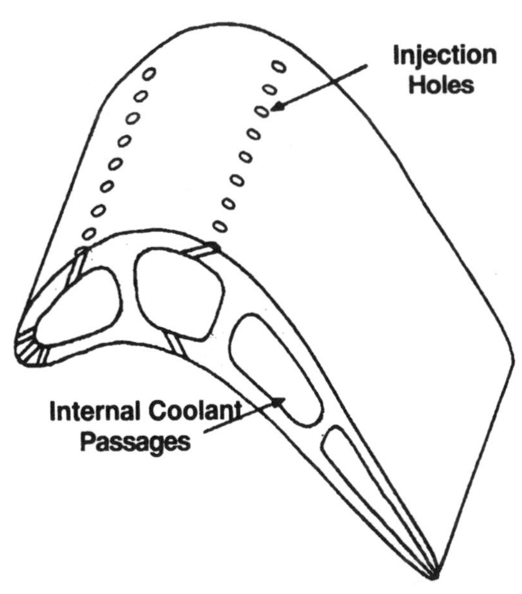

Figure 13-2. Gas turbine blade with film cooling. A coolant such as air travels through the internal coolant passages and comes out through the injection holes to cool the metal blade.

> The effectiveness of film cooling depends on a large number of variables including the hole geometry, pressure gradient, blade curvature, and freestream turbulence level. To design turbine blades that use film cooling, manufacturers have to predict the convective heat transfer in the turbulent boundary layers on these blades.
>
> Typical freestream turbulence levels that occur over turbine blades have been measured by Koutmos and McGuirk [1989] to be greater than 20 percent. All of the literature studies, however, on the effects of freestream turbulence on heat transfer have been done with turbulence levels less than 7 percent. Why? The high freestream turbulence levels of interest (20 percent) cannot be generated and sustained by standard techniques for generating turbulence. The only study that has achieved turbulence levels greater than 20 percent was done by Maciejewski and Moffat [1989]. Recognizing the need for any data that would indicate the effects of high freestream turbulence on heat transfer, Maciejewski and Moffat placed a test plate along the edge of a turbulent free jet. While this experimental condition had the advantage of producing high freestream turbulence, it had definite drawbacks in that the mean velocity field was non-uniform and decayed rapidly along the test plate. Ideally, what is needed is a device for producing high freestream turbulence that has a uniform mean velocity field and that does not decay rapidly downstream.

After presenting this problem, the writers were in an excellent position to propose their work: a new design for a turbulence generator that achieved high turbulence (above 20 percent), but maintained a uniform velocity field.

Proposed Solution. While the problem statement shows the audience that a problem exists and that the writer understands that problem, the second part of a proposal (the proposed solution) presents the audience with a plan for addressing the problem. When well-written, this section answers the questions that the audience has after reading the statement of the problem. These questions depend on the type of proposal. For instance in a research proposal to a funding agency such as the

National Science Foundation, typical questions after the problem statement would be (1) what is the proposed solution? (2) does the solution make sense from a technical perspective? (3) does the solution make sense from a management perspective? and (4) can the person making the proposal carry out the solution?

In answering the first question, you present the scope and limitations of your solution. What is your solution? What are its principal advantages? In this part of the proposal, you show how your solution addresses the problem raised in the first part of the proposal. For instance, if your solution to remove space debris is a huge foam balloon that acts like a collector satellite [Loftus, 1989], then you should explain the type and amount of debris that this satellite could remove. While the problem statement has created a hole in science that needs addressing, the proposed solution presents a plan that fills that void, much in the same way that a piece fills a slot in a jigsaw puzzle.

In answering the second question, of whether the solution makes sense from a technical perspective, you address the technical audience and discuss how you will solve the problem. In the example of the satellite for collecting orbital debris, you would explain the different stages of the proposed work: design of the collector, testing of the collector, placement of the collector into orbit, and collection and disposal of the debris. In answering this second question, you present the methods that you will use to solve the problem and justify why those methods are chosen.

In answering the third question, of whether the solution makes sense from a management perspective, you discuss again how you will solve the problem, but you discuss the solution to the management audience. For that reason, you discuss issues that concern your management audience: the cost of the solution, the schedule for the solution, the effect of the solution on the environment, and the effect of the solution on the safety and health of those people involved. In presenting this information, you respond to the specific questions that your management audience has: Is the solution worth carrying out? Will the solution be carried out in a timely manner? Will the solution have negative effects?

Another aspect of this question is to discuss the effect that the solution will have on the stated problem. Consider an example from a proposal to predict the heat transfer characteristics of gas turbine blades [Thole, 1991]:

> The proposed experiments will help us to predict the heat transfer coefficients for the external surfaces of gas turbine blades. By predicting the heat transfer coefficients, and ultimately the blade temperatures, we can through computer modelling and experimental testing improve the design of blade cooling. This improvement can have a dramatic effect on blade life. For instance, reducing the blade temperature from 1140K to 1090K increases the blade life from 560 hours to 3900 hours [Cohen and others, 1986].

Depending on the situation, the effect on the stated problem can have multiple levels of depth. For instance, in the proposal for a satellite collector of orbital space debris, the effect would certainly include how much orbital debris the collector will remove. A deeper discussion, though, would show how this collection will affect the space program. For instance, how much safer will future space missions be because of the orbital debris removed by the collector?

Finally, in answering the fourth question, of whether the person(s) making the proposal can carry out the solution, you show your readers that you are qualified for the task. Here, you present evidence for your qualifications, your experience, and your resources. Both the management and technical audiences review this information to determine whether you can, in fact, be trusted to carry out the work. Often accompanying this section of a proposal is a résumé in which you summarize your education, experience, abilities, and accomplishments. How you order the categories of a résumé and how much space you devote to each category depend on your background and the expectations of the audience. For instance, if the proposed solution includes experimental tests, presenting your experience on relevant experimental techniques becomes important.

Given that résumés are read quickly, you should format your résumé so that your outstanding characteristics are easy to locate. One way is through vertical lists, which provide a good means for highlighting parallel items that are read quickly.

So far, this section has considered only one type of proposal: a general research proposal to a funding agency such as the National Science Foundation. In this proposal situation, there is not just one problem to be solved, but a scientific area in which many problems exist. For that reason, the submitted proposals do not compete directly with one another on the same problem. Moreover, the reviewers usually focus on finding the most important problems that have acceptable solutions.

What about the situation in which the proposals do compete directly with one another on the same problem? Here, the reviewers focus on finding the best solution to the stated problem. Therefore, as a writer of such a proposal, you have to consider the solutions of your competition when shaping your arguments. You also have to anticipate challenges that your competition will raise about your own solution.

As an example, consider the arguments that Texas made in its proposal to have the superconducting supercollider located in the state. In its proposal, Texas first developed a set of logical criteria for the selection of such a site. These criteria included choosing a site with no coastal sedimentation, no large seismic hazard, no high relief or mountains, no more than 48 inches of annual rainfall, and no more than 120 freezing days per year [State of Texas, 1985]. In this discussion, Texas showed a series of maps that shaded those portions of the country that failed to meet the criteria. At the end of the discussion, the map had a single area left unshaded—an area within Texas. This strategy did two things: (1) accented the advantages of locating the site in Texas and (2) raised arguments against locating the site in other states.

References

Baker, H. A., *Space Debris: Legal and Policy Implications* (Dordrecht, The Netherlands: Martinus Nijhoff Publishers, 1989).

Byworth, S., "Design and Development of High-Temperature Turbines," *Rolls-Royce Magazine*, vol. 44 (March 1986).

Cohen, H., F. G. Rogers, and H. I. Saravanamuttoo, *Gas Turbine Theory*, 3rd edition (New York: Longman Scientific and Technical, 1987).

DeMeis, R., "Cleaning Up Our Space Act," *Aerospace America*, vol. 27, no. 2 (February 1987), pp. 10–11.

Johnson, N. L., and D. S. McKnight, *Artificial Space Debris* (Malabar, Florida: Orbit Book Company, Inc., 1987).

Koutmos, P. and J. J. McGuirk, "Isothermal Flow in a Gas Turbine Combustor—a Benchmark Experimental Study," *Experiments in Fluids*, vol. 7 (1989), p. 344.

Loftus Jr., J. P., (ed.), "Orbital Debris From Upper-Stage Breakup," *Progress in Astronautics and Aeronautics*, vol. 121 (1989), p.12.

Maciejewski, Paul K. and Robert J. Moffat, "Effects of Very High Turbulence on Heat Transfer," *Seventh Symposium on Turbulent Shear Flows* (Palo Alto, CA: Stanford University, August 21–23, 1989).

Nolte, Carl, "800,000 People Trample the Organizers' Plans," *San Francisco Chronicle* (May 25, 1987) p. 6, col. 1.

Pauli, Wolfgang, quoted by H. Coblaus. Original text: *"Ich habe nichts dagegen wenn Sie langsam denken, Herr Doktor, aber ich habe etwas dagegen wenn Sie rascher publizieren als denken."*

State of Texas, *A Proposed Site for the Superconducting Supercollider* (Amarillo, Texas: Texas State Railroad Commission, 1985).

Subcommittee on Space Science and Applications, 100th Cong., 2nd Sess., *Orbital Space Debris* (Washington, DC: U.S. Government Printing Office, 1988).

Thole, K., D. Bogard, and J. Whan-Tong, "A Proposal to Validate a Novel Turbulence Generator for the Study of Turbulent Transport of Heat" (Austin, TX: The University of Texas, 1991).

U.S. Congress, Office of Technology Assessment, *Orbiting Debris: A Space Environmental Problem—Background Paper*, OTA-BP-ISC-72 (Washington, DC: U.S. Government Printing Office, September 1990).

Wright, A. K., "Proposal to Research the Effects of Orbital Debris on the NASA Space Program," submitted to the *Undergraduate Engineering Review* (Austin: University of Texas, 1991).

Chapter 14

Writing Instructions

The disastrous charge of the Light Brigade at Balaclava in the Crimean War was made because of a carelessly worded order to 'charge for the guns'—meaning that some British guns which were in an exposed position should be hauled out of reach of the enemy, not that the Russian batteries should be charged. But even in the calmest times it is often difficult to compose an English sentence that cannot possibly be misunderstood.

—Robert Graves and Alan Hodge

Instructions teach people how to perform processes such as machining a turbine blade, running a computer program, or handling a toxic chemical. Each of these examples points to the importance of well-written instructions. If the specifications for the turbine blade are unclear, the engineer will have to resubmit the job. If the software manual is disorganized, users of the program will waste time searching for commands. If the safety precautions for handling the toxic chemical are ambiguous, someone might become ill. Money, time, and health depend on the quality of the writing in instructions.

Constraints of Instructions

Two constraints that warrant special consideration for instructions are audience and format. Audience warrants special consideration because of why instructions are read and how instructions are read.

The first question, of why instructions are read, is easy to answer. Audiences read instructions to find out how to perform a process, such as using a power supply or running a computer program or meeting specifications for a turbine blade. Unlike an article or report, in which the audience is also interested in what causes the process to occur, instructions are written for audiences who focus on the single-minded, often impatient, question of how the process occurs. For that reason, in formulating your strategy for instructions, you want to anticipate the information that the reader desires and make that information easy to find. For instance, because computer users search for single commands in software manuals, you should provide detailed tables of contents and indexes in those manuals to help users find the commands as quickly as possible. For a manual on handling a toxic chemical, you should include medical procedures on what to do should someone ingest the chemical. In the specifications for producing a turbine blade, you should include several perspectives of the blade to show machinists how the end product will appear.

The one exception to helping the reader quickly find the desired information occurs when a dangerous or difficult step stands between the reader and that information. In such a case, you want to insure that the search for the information passes through the caution for that dangerous or difficult step. Often, that insurance entails placing the caution in glaring, not-to-be-missed typography.

Answering the second question, of how an audience reads instructions, reveals why instructions have such unusual formats. When reading instructions, audiences rarely sit in soft chairs by fireplaces. Often they try to follow the instructions as they are reading the instructions. For that reason, the formats are tailored to make it easier for the audience to read and work simultaneously. For instance, formats for instructions often include numbered steps separated by white space so that the audience can read the step and then turn away from the page and perform the process. The numbers and the white space allow the readers to find their places.

Besides having numbered steps, formats for instructions vary in several other ways. For instance, the lengths of instructions range from single-phrase cautions on tools to thick handbooks on procedures in submarines. Instructions also have unusual typography. In the middle of instructions, you're likely to see red boxes containing cautions. Moreover, the layouts of instructions do not look like the layouts of other documents. There are more illustrations per page in instructions than in other documents. For instance, when reading an owner's manual, you're likely to see the text supplementing the illustrations, as opposed to the reverse case. Layouts for instructions also have more white space than other documents. This additional white space arises from the lists, cautions, and illustrations.

Style of Instructions

Just as the constraint of audience in instructions deeply affects the format, the constraint of audience deeply affects the style. This section discusses the unusual aspects of structure, language, and illustration within instructions.

In a formal set of instructions, the most important structural consideration is how to organize the information into a beginning, middle, and ending. The beginning, which prepares readers to understand the steps of the process, often includes a title, summary, and introduction. In the title, you want to do two things: (1) indicate that the document is a set of instructions, and (2) indicate what process the document will explain. To accomplish the first task, you either format the title as a participial phrase ("Using a Hot Wire Probe") or use key words such as "How to" or "Instructions for." To accomplish the second task, you include the name of the process being presented: "Preparing for an Earthquake" or "How to Locate Metal Fatigue in an Airplane Wing."

For a set of instructions, the summary is usually descriptive rather than informative because an informative summary of a process would just be the step-by-step instructions. For many processes, an informative summary would be dangerous because the audience would read the steps of the process without reading the background that explains those steps or the cautions that precede those steps.

The introduction to a set of instructions answers explicitly or implicitly four questions that readers expect to have answered before they reach the step-by-step instructions: (1) what the process is, (2) why the process is important, (3) what is needed to perform the process, and (4) how the process will be explained. In some instances, you might not have to answer all of the questions explicitly, because the answer is implicitly known by the audience for that process. Also, the order in which these questions are answered depends on the process and the situation.

For an example of a beginning to a set of instructions, consider a document explaining how to play the game three-dimensional tic-tac-toe. As with many scientific instructions, the principal challenge to writing the instructions for this game is to communicate a complex image.

How to Play Three-Dimensional Tic-Tac-Toe

Summary

These instructions explain how to play three-dimensional tic-tac-toe, a new game based on childhood tic-tac-toe, but filled with much more strategy. These instructions explain the game board, give the rules for play, and show winning strategies.

Introduction

What Is Three-Dimensional Tic-Tac-Toe? Three-dimensional tic-tac-toe is a game of strategy based on traditional tic-tac-toe. Like traditional tic-tac-toe, the object of three-dimensional tic-tac-toe is to place marks in either a row, column, or diagonal that spans the edges of the game board. Unlike traditional tic-tac-toe, this new game requires players to imagine the game in three dimensions, thus increasing the game's complexity and increasing your geometric skills.

Who Can Play? Anyone who can imagine a four-by-four-by-four cube can play three-dimensional tic-tac-toe.

What Is Needed to Play? Besides having two players, you need the following materials to play three-dimensional tic-tac-toe:

- the game board
- two marking pens
- a white cleaning cloth.

How Do You View the Game Board? The key to playing three-dimensional tic-tac-toe is to understand the game board. Although the game board appears intimidating, you can understand it if you methodically compare it to a four-planed cube. The game board, which is shown in Figure 14-1a, is nothing more than a two-dimensional representation of a four-planed cube, which is shown in Figure 14-1b. The game board has sixty-four wedges that correspond to sixty-four points on the cube. Each of the sixty-four wedges on the game board appears in one of four quadrants (I, II, III, IV). These quadrants correspond to the cube's four levels. For instance, Quadrant IV corresponds to the cube's bottom level.

To play three-dimensional tic-tac-toe, you have to understand how the wedges on the game board correspond to the points on the cube. Something that will help is to assign rows and columns to the cube that correspond to circles and angles on the game board. As shown in Figure 14-1, rows 1–4 correspond to concentric circles 1–4. Likewise columns *a–d* on the cube correspond to angular divisions *a–d* on the game board. For example, the column *a* of Quadrant I is at 0°, column *a* of Quadrant II is at 90°, column *a* of Quadrant III is at 180°, and column *a* of Quadrant IV is at 270°.

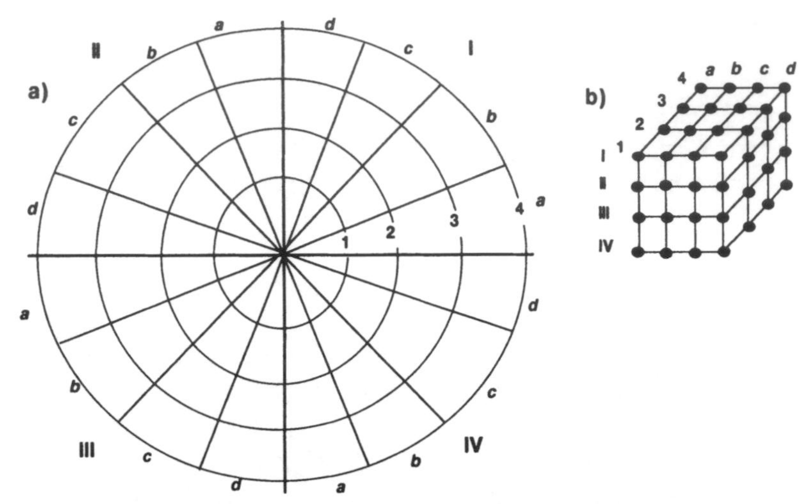

Figure 14-1. a) Game board for three-dimensional tic-tac-toe, and b) corresponding cube that the game board represents. The sixty-four wedges on the game board correspond to the sixty-four points on the cube.

In the middle of a set of instructions, you present the information needed to perform the process. Some processes, such as changing a tire, lend themselves to a single step-by-step explanation. Other processes, such as running a computer program, involve a series of steps, some of them performed by all readers, many of them performed by only some of the readers. Whatever the process, you have to present a logical explanation. A step-by-step explanation is often used for processes that directly depend on time. In this explanation, you start at the beginning of the process and continue through to its end. Some processes, such as using a spreadsheet computer program, do not lend themselves to step-by-step chronology. In those instances, you divide the process into logical sections and subsections. Whatever overall strategy you use, you should provide enough detail that your audience can perform the process. The following middle continues our tic-tac-toe example:

Step-by-Step Instructions

Setting up the Game. To prepare to play three-dimensional tic-tac-toe, do the following:

1. Take out the game board and lay it flat.
2. Have each player choose a marking pen.
3. Flip a coin to see who moves first.

Playing the Game. To play three-dimensional tic-tac-toe, do the following:

1. Have the first player draw an X in any of the sixty-four available wedges of the game board.
2. Have the second player draw an O in any of the remaining sixty-three wedges of the game board.
3. Play continues with players alternating turns until one player occupies four wedges in a row (either across, up, down, or diagonally through the imaginary cube) *and* declares a win.

Note: a player (say X) who occupies four in a row but does not declare a win on that turn must wait until the next turn before declaring the win. Should the other player (O) then occupy four in a row and declare a win, then O is the winner.

4. In the unlikely event that player X and player O mark all sixty-four wedges with no one securing a victory, then the game is a tie (called a "cat").

Determining a Winner. Unlike childhood tic-tac-toe, in which there are only eight possible winning combinations, three-dimensional tic-tac-toe has seventy-six possible winning combinations. Such a combination might lie entirely in one level (quadrant) or extend through all four levels (quadrants).

Winning combinations on one level can be angular, radial, or diagonal. Figure 14-2 presents an example of each in Quadrant II. Figure 14-2a is an example of an angular win in Quadrant II, Figure 14-2b is an example of a radial win in Quadrant II, and Figure 14-2c is an example of a diagonal win across the Quadrant II. For each quadrant there are ten winning combinations.

For wins that span all four levels (quadrants), there are three types: wins straight down through the cube; diagonals along one of the eight vertical planes of cube; and diagonals linking opposite corners of the cube. An example of each is shown in Figure 14-3. Note that there are sixteen possible wins straight down through the cube, sixteen possible wins that go diagonally along the vertical faces of the cube, and four possible wins that link opposite corners of the cube.

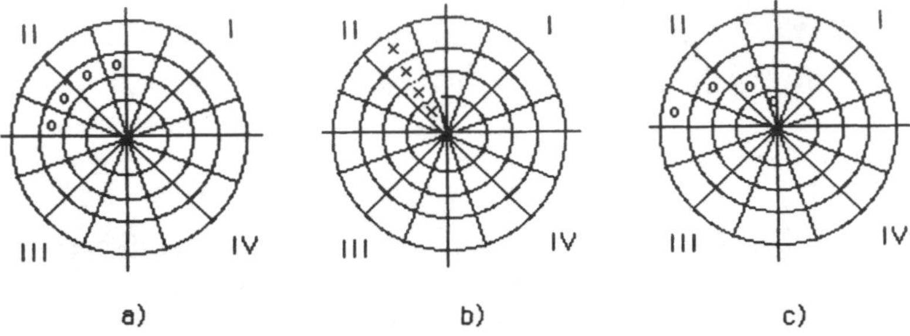

Figure 14-2. Examples of winning combinations on a single level (in this case, Quadrant II): a) angular win; b) radial win; c) diagonal win.

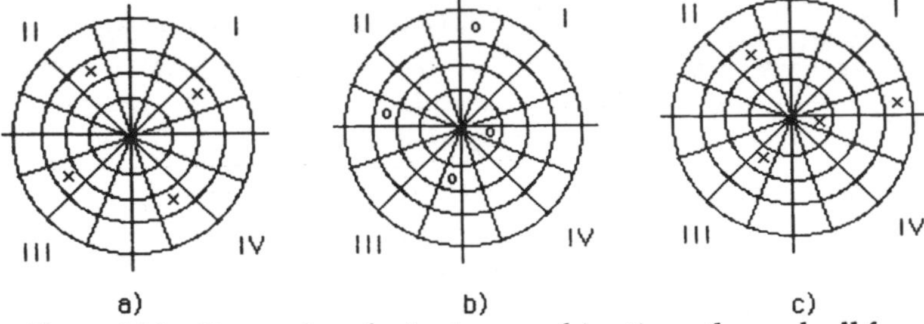

Figure 14-3. Examples of winning combinations through all four levels (or quadrants) of the cube: a) straight down through the cube; b) diagonal along a vertical plane of the cube; and c) diagonal connecting opposite corners of the cube.

Developing Winning Strategies. There are numerous strategies that you can develop for winning three-dimensional tic-tac-toe. Most of these involve setting up a trap in which the opponent is left with only one move to counter two or more possible wins. One strategy is to saturate one plane and set up a trap in that plane. Another strategy is to control the center blocks of the cube (found in Quadrants II and III) and set up a trap that spans the four levels of the cube.

Be careful, though, about concentrating too much on setting up a trap. To win three-dimensional tic-tac-toe, you must also be aware of your opponent's moves.

The ending to a set of instructions usually does two things: (1) summarizes the process by showing the interrelation of steps, and (2) gives a future perspective on the process. Unlike the instruction set's initial summary, the summary in the conclusion is addressed to an audience that has already read the instructions (including the cautions). This summary, therefore, can take more liberties than the initial summary, which for instructions is usually descriptive.

The second thing that an ending to a set of instructions does is to give a future perspective to the process. In a computer software manual, that perspective might include such things as an index, so readers can reference specific problems. In an instructional brochure on how to operate a streak camera, that perspective might include a troubleshooting guide. As an example, consider the ending to our instructions for three-dimensional tic-tac-toe:

Conclusion

Summary of Game. Three-dimensional tic-tac-toe is played much the same as traditional tic-tac-toe, except that to win you must get four, rather than three, marks in a row, column, or diagonal. Also, unlike traditional tic-tac-toe, which has only eight winning combinations, three-dimensional tic-tac-toe has sixty-four possible winning combinations, thereby providing much more variety and strategy. Finally, unlike traditional tic-tac-toe, which is played in only two-dimensions, three-dimensional tic-tac-toc challenges your spatial skills because it is played in three dimensions.

Variations on Game. Because of its high number of winning combinations, three-dimensional tic-tac-toe can be played in various ways. For instance, you might want to play with three players, rather than two. The third person could use a Δ for his or her symbol. Play would proceed the same as before, except that three players would rotate turns rather than two.

Another variation is to play until all sixty-four wedges are filled and then count the number of wins for each player. The player with the most number of wins is the winner.

The language of instructions is quite different from the language of other types of documents. For instance, instructions often include sentences written in the second person (with the word "you"). Moreover, instructions include sentences written in the *imperative mood*. In a sentence written in the imperative mood, the subject is an understood "you" (for example, "During an earthquake, stand in a doorway or crouch beneath a table or desk"). In most instructions, you do not use the imperative mood in every sentence; rather, you use it for important steps.

Another interesting language aspect of instructions is the use of cautions, often in highlighted type, to warn readers of difficult or dangerous steps:

> *Caution: Should an earthquake cause the power to go out in your building, do not light a match. Gas lines often break during earthquakes, and lighting a match can trigger an explosion.*

Because the emphasis in instructions is telling someone "how," being clear becomes the most important goal. In instructions, being clear means not having any ambiguities, especially those that could lead to mistakes or injuries. In the instructions for three-dimensional tic-tac-toe, you wouldn't want to confuse the audience as this writer did:

> To set up three-dimensional tic-tac-toe, take out the game board and lie flat on a hard surface. Now, take out the two marking pens and divide them equally between the two opponents. Now, toss a coin in the air and have one player call either heads or tails. Play continues until someone is declared a winner.

Following this writer's instructions, the players end up flat on their backs, they have broken their marking pens (and probably stained the carpet), and they are tossing a coin indefinitely until an angel of mercy descends and declares one of them a winner.

As stated earlier, instructions typically have more illustrations per page than other kinds of scientific documents. The reason for the increased number of illustrations is that people often learn how to do a process by seeing the process. Developing illustrations for instructions calls on your imaginative skills. The goal is to find perspectives that allow your audience to understand the process. For instance, an imaginative way to illustrate the game board in the tic-tac-toe example is to show how the three-dimensional cube transforms into the two-dimensional game board. Figure 14-4 shows a key step for this transformation. In this step, a face of the cube has collapsed to a line and the four planes along that line have separated into "wedges." To form the game board, all you have to do is rotate each wedge to a quadrant position and collapse the wedges to a single plane.

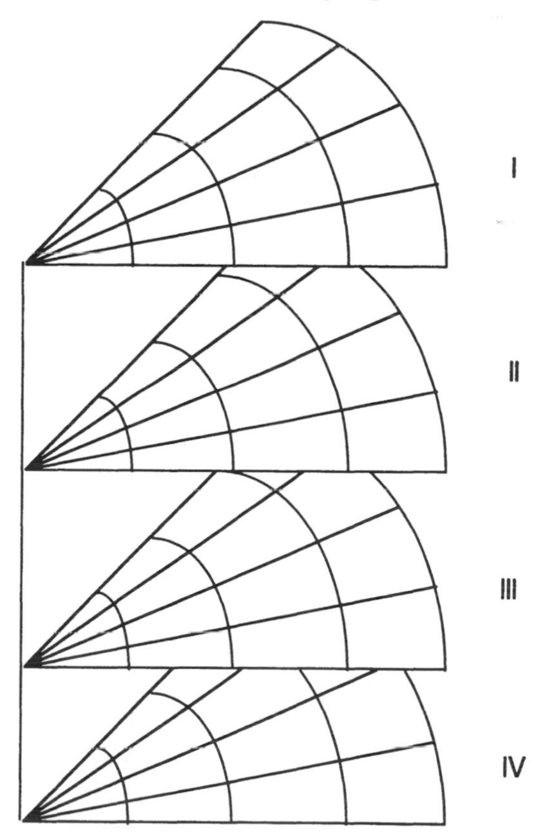

Figure 14-4. A key step for transforming a four-by-four-by-four cube (Figure 14-1b) into the game board (Figure 14-1a) of three-dimensional tic-tac-toe.

Although you should use illustrations generously, each illustration should serve the document. In other words, gratuitous illustrations can confuse the reader in the same way that gratuitous words can. Some processes, such as performing an error analysis, do not lend themselves to illustrations. In such processes, you should use examples to anchor the steps.

Reference

Graves, Robert, and Alan Hodge, *The Use and Abuse of the English Language* (New York: Marlowe & Company, 1995), p. 95.

Chapter 15

Preparing Presentations

On Monday and Wednesday, my mother was nervous and agitated from the time she got up....My mother had been teaching for twenty-five years, yet every time she had to appear in the little amphitheater before twenty or thirty pupils who rose in unison at her entrance she unquestionably had "stage fright."

—Eve Curie

During World War II, Niels Bohr gained an audience with Winston Churchill to warn him about the dangers of atomic weapons. Bohr realized that after the war an atomic weapons race would begin, and he wanted all countries to establish guidelines for containing these weapons. Despite the importance of the information that Bohr had to convey, Churchill broke off the meeting after only fifteen minutes. The reason? Bohr was such a poor speaker that Churchill lost patience trying to understand him.

Bohr's failure as a scientific presenter is not unusual. How many times have you found yourself in the middle of a scientific presentation, eager to learn about the topic, yet confused or frustrated or, worse yet, bored? In science and engineering, presentations are essential for conveying information. Yet all too often, presentations in science and engineering fail to inform. All too often, they are needlessly complex, they fail to accentuate important facts, and they lack the enthusiasm needed to maintain audience interest.

You might ask, "Why even make a presentation? Why not just present the information in a document and then distribute it?" In some situations, a document would better serve the audience than a presentation. For that reason, before preparing or even scheduling a presentation, you should weigh the advantages and disadvantages of a presentation in relation to a document.

Presentations offer several advantages over documents. First, in a presentation you've got someone on stage to make the work come alive. For instance, reading an article about fossil-dating is much less exciting than seeing the archaeologist describe how he or she found and dated the fossil. Second, in a presentation, the speaker has the opportunity to observe the audience and revise on the spot. During a presentation to some mathematicians, Patrick McMurtry, an engineering professor from the University of Utah, noticed from the blank looks of his listeners that they did not understand the term "laminar steady-state flame." McMurtry asked to borrow someone's lighter, clicked it on, and gave the audience an example. Because this term was crucial to understanding the work, McMurtry salvaged the presentation with this on-the-spot revision.

Besides having someone make the work come alive and the opportunity to revise on the spot, presentations offer many exciting ways to present information that are unavailable in a document. For instance, in a presentation you can present films, demonstrations, and three-dimensional models. You can also incorporate color more easily into a presentation than you can into a document.

What are the disadvantages of a presentation? First, while you have many opportunities to revise in a document, you have only one chance to say things correctly in a presentation. Simply forgetting a word from a sentence in a presentation can trip your audience, especially if that word is important—the word "not," for example. Likewise, in a presentation, your audience has only one chance to hear what you say. If someone daydreams for a moment during your presentation, that person misses what you have said. A document, on the other hand, allows readers to review a passage as many times as they desire. Finally, in a presentation, the audience has no chance to look up background information. If in a presentation you use an unfamiliar word, such as "remanence," and do not define the word, then your audience is stuck. If the presentation's format does not allow for questions during your presentation, then members of the audience sit frustrated wondering what "remanence" means. With a document, though, the reader has the chance to look up "remanence," which is the residual magnetic flux density in a substance when the magnetic field strength returns to zero.

Constraints of Presentations

As with documents, the way you put together a presentation depends on your constraints—in particular, the constraints of audience and format. With audience, you have essentially the same considerations as you do in documents, with one important difference. That difference becomes apparent when you face a multiple audience, such as an audience that includes scientists, engineers, and managers. While in the report you would write the text for the primary audience and use the back matter to relay information to the secondary audience, you do not have that option in a presentation. Rather, you have to find a path through your work that addresses the entire audience.

With format, things change even more dramatically. In a presentation, format has a different meaning than it does in a document. While the format of a document refers to the document's typography and layout, the format of a presentation refers to other things such as the length of the presentation, the time of day that the presentation occurs, and the available equipment for projection. Format can substantially affect how you tailor your presentation. For example, if you are deciding whether to present three or four main points, you might choose four points for a presentation at ten o'clock in the morning, when you can expect a lively audience, but only three main points for a presentation at four o'clock in the afternoon, when you can expect a tired audience. Also, if the formality does not allow for questions until the end of the presentation, you should be even more sensitive about defining terms and providing background than you would be if the audience has the opportunity to ask questions throughout the presentation.

Style of Presentations

Because presentations have advantages and disadvantages in regard to a document, you should tailor the style of your presentations to accent the advantages and to mitigate the disadvantages. This section discusses some ways to tailor the style of presentations by examining presentations from a writing perspective. Much writing goes into a presentation, particularly in developing the presentation's organization and creating the presentation's visuals. This section also examines presentation delivery, a perspective that has no parallel in writing.

Organizing Presentations. In organizing a presentation, you should think of the presentation in terms of its beginning, middle, and ending. Much about the organization of a presentation is similar to the organization of an article or report. Still, because of the intrinsic advantages and disadvantages of a presentation, certain aspects of organization are more important than others.

What is most important about the beginning? To answer this question, put yourself in the position of the audience before a presentation begins. Imagine yourself sitting in the room. Your chair is comfortable, but you are restless. You are a busy person who has to finish a set of experiments or run a computer simulation, and yet you are here waiting for someone to give a presentation. A few days earlier, the presenter's title and summary interested you, but now the specific details seem cloudy. All you can think about are your own experiments or simulations. The presentation begins. The presenter moves to the front of the room, a hush falls over the crowd, and you begin to wonder (1) what this presentation is really about, (2) why this presentation is important, (3) whether you will understand this presentation, and (4) how this presentation will be arranged.

As with a document, these four questions loom in the beginning of the presentation. Likewise, as in the beginning of a document, some of these questions might be answered implicitly, given the particular audience and subject matter. Nonetheless, after the beginning, no one in the audience should have any of these four questions lingering in his or her mind.

The last of these questions is more important in a presentation than in a document. Why? Unlike a document, in which the readers can glance ahead to see the headings and subheadings and therefore see what information will occur, the listeners to a presentation have no idea where the presentation is going unless the presenter tells them. Note that in answering this question of how the details will be presented, the presenter maps the organization for the entire presentation. When the presenter clearly maps the organization, the audience is aware at any point in the presentation about how much has been covered and how much further the presenter has to go. That information is important because listeners pace themselves. Listening is hard work, and asking someone to listen, especially to a scientific presentation, without giving a clue as to the path of that presentation, is similar to taking that person on a snipe hunt. Because the person doesn't know how far he or she is going, the person quickly tires.

In the middle of a presentation, your organizational goal is the same as in a document: find a logical path to explain your work to your audience. Sometimes that path is chronological. Other times that path follows a flow of energy or mass or information through a system. On still other occasions, you divide the path into parallel parts—perhaps three different drugs to battle Alzheimer's disease. Whatever path you choose, the path should have logical divisions. Most important for a presentation, the number of divisions in the path should be low. People remember clusters of twos, threes, and fours. Clusters larger than four tax the listener.

The conclusion of a scientific presentation is as important as the beginning. As with the beginning, the ending is an occasion when people sit up and concentrate. In a presentation, people remember best what is presented first and then what is presented last. While you use your first part to orient your audience, you use the last part of a presentation to restate your most important ideas. In the conclusion of a presentation, you accomplish the same two tasks as in the conclusion of a document. You present an overall analysis of what occurred in the middle, and you give a future perspective.

Creating Visuals for Presentations. A second important role that writing plays in presentations is in the creation of visuals (slides or transparencies). Why use visuals? One reason is that many images in science and engineering are just too complex to communicate with only words:

the magnetic field lines at the poles of the earth, the life-stages of a Hawaiian volcano, the markings of a peacock. Another reason is that people remember images much more easily than they remember words. Close your eyes for a few moments and think about your first year in grade school. What comes back to you? You're more likely to recall silent scenes—a misty morning waiting for the bus or the blocked letters of the alphabet above the blackboard—than to remember exact words anyone spoke. If you do recall exact words, you probably also recall the images from the scene in which those words were spoken.

Besides providing images, visuals can reinforce the structure by emphasizing results and smoothing transitions. Finally, visuals can give your presentations a unique look; they can offer variety and beauty to your presentations. In a way, visuals act as another presenter. For many of us, though, designing visuals is an intimidating venture. Although we are professionals in the laboratory or computer room, we feel like apprentices when wielding a camera or a graphics computer program. Fortunately, most companies, laboratories, and universities have professionals to assist us. Still we waver. After all, allowing someone else partial control of our presentation is frightening. How will they know to choose the best perspective for the photograph? What if they misspell "velocimetry" on a call-out? The answer is that you've got to work with these people. You should cultivate strong (and lasting) relationships with the support professionals where you work.

When designing visuals, you should consider their effect on the audience. For instance, when you place a visual on the projector, the listeners break their eye contact with you and look at the screen. They try to discern what is projected and how this projection fits into the scheme of the presentation. When visuals work well, the visuals orient the audience quickly.

What information do you place on visuals? First, you

should use visuals to show the presentation's organization. Figure 15-1 presents the mapping visual from a presentation on ways to reduce sulfur dioxide emissions from coal-fired power plants. This visual introduces the three categories of methods that will be discussed in the presentation. Note that this mapping visual does much more than just list the three categories of methods. This visual depicts the process for bringing the coal to the plant, burning it, and dispensing with the emissions. These images provide the speaker with many opportunities to work in background information.

Besides including the presentation's organization on your visuals, you should also use your visuals to give important results from the presentation. For instance, Figure 15-2 presents the effectiveness of one of the methods mapped in Figure 15-1. In Figure 15-2, an image of the process and four key results are given. Note that both visu-

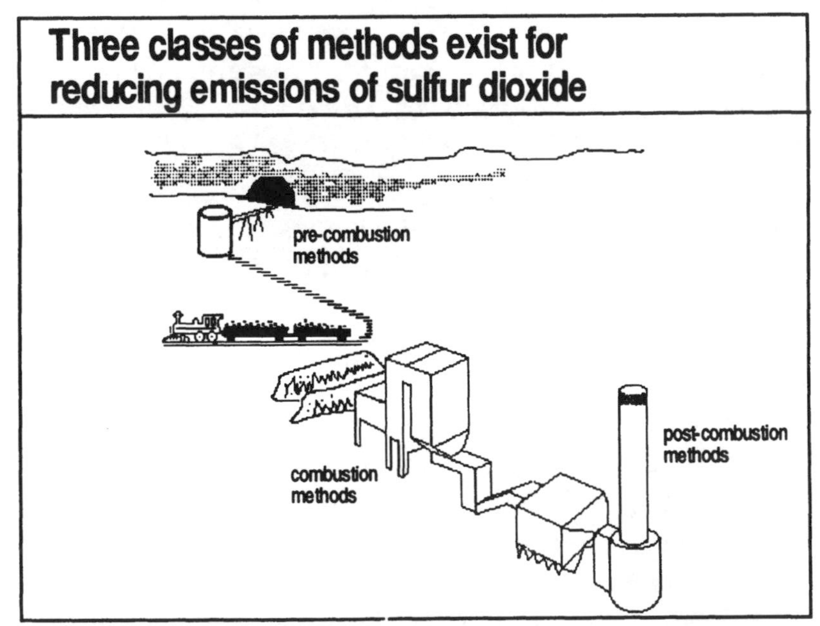

Figure 15-1. Visual that reveals the organization of a presentation on reducing sulfur dioxide emissions from coal-fired utilities [Schmidt, 1989].

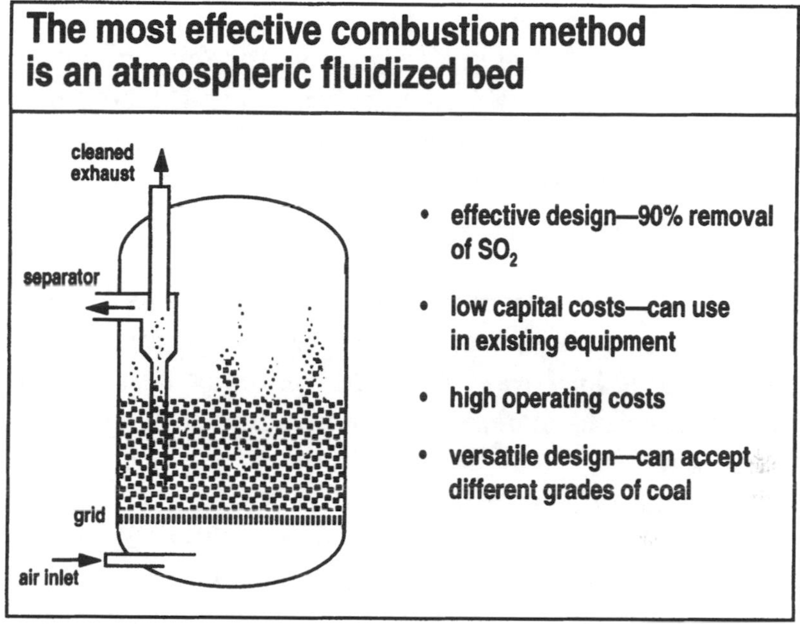

Figure 15-2. Visual that presents the key results of a presentation on reducing sulfur dioxide emissions from coal-fired utilities [Schmidt, 1989].

als (Figures 15-1 and 15-2) have a nice balance of images and words. After the presentation, people will remember first the images of the visual, and the memory of the images will help trigger the memory of the wording. Because people remember images, you should try to include an image with each visual.

In addition to having a nice balance of words and images, each visual has a headline at the top that orients the audience to the purpose of the visual. This headline is a sentence, rather than a phrase, because a sentence can clarify much better than a phrase can what the perspective of the topic is. For instance, the sentence headline for the visual of Figure 15-2 delivers much more information than the phrase "atmospheric fluidized bed." Many professional presenters such as Larry Gottlieb [1985] at Lawrence Livermore National Laboratory feel that you should include a sentence headline with each visual. If the visual does not make a point that can be stated in a sentence headline, then that visual is probably not worth including. When writing a headline, keep the sentence short (no longer than two lines) so that the audience can quickly read it and focus attention back on you.

Many visuals projected at scientific and engineering presentations communicate poorly. Overcrowding is perhaps the most common weakness. Symptoms of overcrowding include lack of white space, long passages of text, and long lists (more than four items). When a visual has too much information, the audience often doesn't bother trying to read it.

Besides overcrowding, another common mistake is having lettering that is too small. In general, the typesizes should be between 18 and 36 *points*. At a recent national conference, in which presenter after presenter used 10 and 12 point type on their visuals—a size that people sitting in only the first couple of rows could read—one person in the audience decided that he had had enough. This person moved to the back of the auditorium, stood on a chair, and focused a pair of binoculars onto the screen. Because most of the audience had long since given up trying to read the tiny lettering on the screen, they soon spotted the man in the back with the binoculars. A wave of laughter and finger pointing passed over the auditorium. The commotion was so loud that the presenter became flustered and turned off the projector. For this presenter, I have no sympathy. Not taking the time to create a visual that the entire audience can read is inconsiderate.

Another common mistake in designing visuals is not achieving a balance between the words you say and the words you show. In a strong presentation, you often repeat words and phrases from the visuals in your speech, but your speech should include more than the words on the visuals—much more. In many weak presentations, all the words that the speaker says are given on the visuals. Such a presentation disturbs an audience. The audience isn't sure whether to listen or to read. On one occasion, the Secretary of Energy visited a national laboratory, where he heard a presentation given by a department manager. The manager had worked for weeks on this presentation. The manager booked the best conference room at the lab, he had the best artists at the lab design the visuals, and he practiced the presentation over and over until he could say every word on the visuals without even looking at the visuals. After the third visual, the Secretary of Energy raised his hand. The department manager stopped and said, "Yes, you have a question?"

"No. No, I don't have a question," the Secretary of Energy said. "I have a comment. I can read. From now on, don't say anything else. Just put the visuals up one by one. I'll tell you when to change them."

As you might imagine, the department manager was humiliated. For this department manager, I do have sympathy. Given the work that he put into his presentation, he deserved better than to be belittled. Still, the story points to the importance of achieving a balance in what you say and what you show. Your visuals should certainly include important images, important results, and key words that show the organization of your presentation. However, your speech should have things not found on the visuals, such as minor results, examples, anecdotes, and background information.

Delivering Presentations. In a presentation, delivery is the way you actually communicate to the audience. Delivery includes such things as your voice, gestures, and posture. As much as organizing the details and creating visuals, delivery determines the success of a presentation.

One important aspect of delivery is the source for your spoken words. Do you memorize your speech, do you speak from a prepared text, or do you speak from outlined points? In general, you are more effective when you speak from outlined points because speaking from outlined points forces you to use natural wording and allows you to keep eye contact with the audience. Speaking from outlined points does not mean that you just jot down a couple of phrases and then "wing it." Speaking from points requires a lot of preparation and practice.

Another important aspect of delivery is your stage presence. As a speaker, you should try to convey enthusiasm for your subject. In my experience, the most captivating speakers in science and engineering have been the ones who loved their subjects and knew them well. If you do not convey your interest for the subject, how can you expect your audience to become interested? Conveying enthusiasm does not mean that you sing and shout. If you put on pretences, the enthusiasm will not be sincere, and the audience will see through the act. You can remain low-key and still be enthusiastic. What's important is that you funnel your energy into delivering the information to the audience.

Just the act of speaking before an audience generates a lot of nervous energy for presenters. There's nothing unusual about feeling nervous before a presentation. Many great scientists such as Marie Curie and great actors such as Jimmy Stewart have shared these same feelings. Another example is Sir Lawrence Olivier, who claimed to have felt nauseated before every stage appearance. What distinguishes the best speakers is their ability to channel that nervous energy into their presentations. How do you make nervousness work for you? First, you should "think positive." The nervousness that a presenter feels is similar to the nervousness that an athlete feels. How do athletes handle nervousness? Many athletes imagine themselves succeeding. When pitching for the San Francisco Giants, Dave Dravecky imagined the ball finding the spot in the catcher's mitt before he ever released the ball. While waiting to return a service, Jimmy Connors imagined not only hitting the ball but also the flight of his service return across the net. In his book *The Inner Game of Tennis,* Tim Galloway [1972] presents an excellent discussion on the power of positive thinking. Before your presentation, imagine yourself delivering a successful presentation. Imagine yourself delivering all your points and making smooth transitions between all your visuals.

Too much nervousness can mean that you're not prepared for the presentation. Before an important presentation, you should have at least one practice run. On this practice run, you should use your visuals. If you're unsure about yourself, either videotape the dry run or else have some friends attend. Although you should consider incorporating valid criticism, you should not go overboard and make major changes right before the presentation. Too many changes just before a presentation may cause more harm than good.

If you've done your preparation, then the structure, speech, and visuals of your presentation should be ready. What's left are the little things that take place in the presentation room that smooth your delivery: making sure the projector is focused, making sure your visuals are in order, making sure the lights are as you want them. Once you've set things up in the room, you should concentrate on your listeners. Meet them before the presentation and ask them questions. By concentrating on your listeners, you shift your thoughts (and worries) away from yourself and give needed attention to your audience. Remember: You're working for them. If you can focus your attention on your audience, then any residual nervous energy is going to work for the presentation, not against it.

Many people worry too much about all the different aspects of delivery: gesture, stance, eye contact, and so on. An alternative approach is to think about delivery from an overall perspective. One way, for example, is to imagine someone who is an excellent speaker making your presentation. Imagine the rhythm of his or her voice. Imagine his or her movements on stage. And then make your presentation. A model speaker for me is Dan Hartley, a Vice-President at Sandia National Laboratories. Hartley is sincere, professional, and acutely sensitive to audience. Having him as a model does not diminish my individuality as a speaker. Rather, it helps me bring out those traits in my own delivery that I value so highly in his.

References

Curie, Eve, *Madame Curie*, trans. by V. Sheean (New York: Da Capo Press, Inc., 1986).

Galloway, T., *The Inner Game of Tennis* (New York: McGraw-Hill, 1972).

Gottlieb, Larry, "Well Organized Ideas Fight Audience Confusion," article (Livermore, CA: Lawrence Livermore National Laboratory, November 1985).

Schmidt, Cynthia M., "Ways to Reduce Sulfur Dioxide Emissions From Coal-Fired Utilities," presentation (Austin, Texas: Department of Mechanical Engineering, 8 December 1989).

Chapter 16

Format: Dressing Documents for Success

Printing should be invisible. Type well-used is invisible as type. The mental eye focuses through type and not upon it, so that any type which has an excess in design, anything that gets in the way of the mental picture to be conveyed, is bad type.

—Beatrice Warde

When you create a document on a computer, you are faced with many formatting decisions, from choosing a typeface to selecting the amount of white space that precedes a heading. In making these decisions, you want to choose a format that is easy to read, that is in character with the kind of document you have written, and that presents the work in such a way that the most important details stand out. Because of the wide selection of typography and layouts, the best thing to do is to consult a graphic designer. Most graphic designers have spent years studying different designs and can tailor a format to meet your needs.

Sometimes you will not have easy access to a graphic designer. In such cases, you will have to select the typography and design the layout yourself. This chapter presents guidelines for selecting typography and designing layouts. Appearing at the end of this chapter is a list of additional references.

Typography of Documents

Typography is that part of format that deals with the choice and size of typestyles. A typestyle (or font) is a shaped set of alphabetic letters. Hundreds of typestyles exist, a few of which appear in Table 13-1.

Table 13-1
Common Typestyles in Scientific Documents

Typestyle	Characters	Type	Uses
Antiqua	abcdefghijklmnop qrstuvwxyz	serif	correspondence, reports
Old English	abcdefghijklmnop qrstuvwxyz	serif	(out of date)
Palatino	abcdefghijklmnop qrstuvwxyz	serif	correspondence, reports, articles
Schoolbook	abcdefghijklmnop qrstuvwxyz	serif	correspondence, reports, books
Times	abcdefghijklmnop qrstuvwxyz	serif	correspondence, articles
Arial Narrow	abcdefghijklmnop qrstuvwxyz	sans serif	illustrations, visuals, headings
Helvetica	abcdefghijklmnop qrstuvwxyz	sans serif	illustrations, visuals, headings
Optima	abcdefghijklmnop qrstuvwxyz	sans serif	illustrations, visuals, headings

The typestyle of a document says a lot about the document. For instance, the font Times conveys a sense of professionalism and authority (Times appears in several newspapers). Times belongs to a class of fonts known as serif fonts, which have projecting short strokes, such as the little feet on a serif "m." Another category of fonts is sans serif fonts, which do not have these projecting strokes (consider a sans serif "m"). One of the most common sans serif fonts is Helvetica.

In making choices for typography, you want to consider the subject matter, occasion, and audience. One font that is appropriate for the window display of a flower boutique may not be appropriate for the text of a report about flower germination. Given below are some general guidelines for typography.

1. *Do not use too many typestyles in a document.* Some people mistakenly try to use all the typestyles on their computers. What occurs in these situations is a mess. In a short document, such as a memo, one typestyle suffices. In a longer document, you might use a serif font such as Times for the text and a sans serif font such as Narrow Helvetica for the headings and illustration call-outs. This combination works well—the sans serif font sets apart the headings and illustration call-outs from the text.

2. *Rely on serif fonts for the texts of documents.* If you look at most books and journals, you will see that publishers have used serif fonts for the text of the documents. For example, many newspapers use a variation of Times, and many textbooks use a variation of a font called Schoolbook. Note that newspapers, which have several columns, often opt for Times because it is narrow. Textbooks are generally single-columned and are better suited for a wider font, such as Schoolbook.

Why not use sans serif fonts for the text of a document? One reason is that sans serif fonts do not have a connected baseline. This baseline makes it easier for the eye to jump from one line to the next, thereby preserving the continuity of the reading. For short lines, such as in a pamphlet, this jumping does not pose a problem. However, for longer lines, such as in a textbook, the reader benefits from the continuity that a serif font provides.

Another reason to use a serif font is tradition. Almost all professional books and journals have serif fonts as the typefaces of their texts. Moreover, because many publishers use sans serif fonts for the texts of grade-school readers, many people associate sans serif fonts with that class of writing. While some sans serif fonts such as Optima are acceptable for the texts of professional documents, many sans serif fonts such as Geneva are not.

3. *Be conservative with options such as boldface or italics.* Too much boldface will overwhelm a page and intimidate a reader. Italics, another typeface option, is also difficult to read in large blocks. Occasions in which boldface is appropriate would be in headings or subheadings. With italics, appropriate uses include subheadings, glossary terms, foreign words, and accented sentences (such as commands in instructions). An underline is a poor person's substitute for italics. If your computer has italics, use it instead. Finally, shy away from cute options such as shadow or outline.

4. *Use a size that is appropriate for the occasion.* Typeface size is measured in points (a point is about 1/72 of an inch). Table 13-2 gives occasions when various sizes of typefaces are used. As a general rule, use 12-point type for the text of documents that are single columned and use 10-point type for the text of documents that have multiple columns. For presentation visuals, use sizes between 18 and 36 points.

5. *Avoid typography that adds needless complexity.* Given the inherent complexity of the subject matter in scientific writing, documents are difficult enough to read. For that reason, you should avoid format guidelines that make the typography even more difficult. For instance, avoid ab-

Table 2
Uses for Different Typeface Sizes

Size	Use
36 points	posters, visuals
24 points	posters, visuals, titles
18 points	visuals, titles, headings
14 points	titles, headings
12 points	text, illustration call-outs
10 points	text, illustration call-outs
<10 points	footnotes

breviations for words such as "figure" or "reference." The additional periods don't save the readers that much and only serve to make the document appear more complicated.

Also, avoid strings of all capital letters. Many presenters mistakenly use all capital letters on their visuals. These presenters fail to recognize that readers recognize words not only by the letters in the word, but also by the shape of the letters—for instance, the shapes of ascenders such as *b*, *d*, and *f* and the shapes of descenders such as *g*, *j*, and *p*. Using all capital letters prevents readers from recognizing the shapes of words and dramatically slows the reading.

If you feel compelled to use something besides boldface or italics to separate a heading or illustration call-out, then consider using SMALL CAPITALS, rather than ALL CAPITALS. Because SMALL CAPITALS take up much less space than ALL CAPITALS, they are softer on the eye.

Layout of Documents

Layout is the arrangement of the words on the page. Layout includes the number of columns, the spacing between lines, and the widths of margins. In general, you should either follow the guidelines suggested by your word processor's default settings or mimic the layout of a professional publication that you find attractive. Below are some general suggestions.

1. *Consider subject matter and audience in layout decisions.* For instance, if the document will have several articles that the audience will read at different sittings, consider a multiple-column layout such as in a newspaper. If the subject matter is a single subject that the audience will read methodically, consider a single-column format such as in a textbook.

2. *Be generous with white space.* White space is needed for margins, column divisions, headings, and illustrations. White space is important because it helps emphasize information and draws readers into the text.

3. *Choose a hierarchy for the headings and subheadings.* Because headings and subheadings show the hierarchy of topics, you should design a format that visually reflects that hierarchy. Two means of achieving that visual hierarchy are white space and typography. The more white space around a heading, the greater emphasis that heading will receive. For that reason, a first-level heading will have more white space around it than a second-level heading. How should you proportion the white space above and below the heading? As with most aspects of format, there is no definitive answer, but a common practice is to

leave twice as much space above a heading as below. With such a proportion, the reader can easily see which block of text belongs with which heading.

Besides using white space, you can use typography to separate headings. One way to designate hierarchy with typography is to vary the typesize. The higher level the heading, the larger the typesize. Another way to separate headings is to use typographical features such as boldface.

Note that in most professional layouts, a combination of white space and typography is used. Consider the following example, which is based on the format of reports at Sandia National Laboratories [1990]. This example uses both typography and white space to give hierarchy:

1st Level:	18 or 14 point, bold; 3 lines before, 2 lines after; centered
2nd Level:	14 or 12 point, bold; 2 lines before, 1 line after; left justified
3rd Level:	12 point, bold; 1 line before, 0 lines after; indented and at beginning of paragraph

In a longer document, a pleasing variation is to use a sans serif font such as Narrow Helvetica for the headings and a serif font such as Times for the text. The sans serif font distinctly sets off the headings from the text.

References

Hwang, A. D., "Writing in the Age of LaTeX," *Notices of the AMS*, vol. 42, no. 8 (August 1985), p. 878.

Parker, *Looking Good in Print* (Chapel Hill, North Carolina: Ventana Press, 1987).

Sandia National Laboratories, *Format Guidelines for Sandia Reports*, SAND90-9001 (Livermore, California: Sandia National Laboratories, 1990).

Shushan, Ronnie, and Don Wright, *Desktop Publishing by Design*, 3rd ed. (Redmond, Washington: Microsoft Press, 1993).

White, Jan, *Graphic Design for the Electronic Age* (New York: Watson-Guptill Publications, 1987).

Chapter 17

Actually Sitting Down to Write

The writing aspect of scientific research is exhausting... I have rewritten many parts of papers four to six times, restructuring the entire organization, before I finally became satisfied.

—Hermann Helmholtz

No one can tell you the most productive way to get your words onto paper. Perhaps you can learn something from imitating the habits of professional writers, but the actual act of sitting down to write is individual. What has worked for Joan Didion or James Agee might not work for you.

Scientific writing is hard work. Granted, it's not as physically exhausting as swinging a pick or as mentally demanding as solving a nonlinear differential equation, but it requires concentration and patience. Moreover, the solutions are not exact. You don't draft a document and sit back and say, "Perfect." No matter how many times you revise a document, there will always be phrases that won't sound right, as well as sentences where you feel compelled to state about ten details at once.

Scientific writing is often lonely work. Although you

might brainstorm ideas in a group or solicit sections of a document from other people, the process of sitting down to write calls for periods of solitary confinement. Just because scientific writing is difficult does not mean it is drudgery—not at all. Writing a scientific document demands energy. You have to convey complex ideas and images. Strong scientific writing also demands imagination. You have to detach yourself from your work and place yourself in the position of your audience. Moreover, you will find that writing a strong document challenges not only your writing skills, but also your scientific and engineering skills. When you write to inform or to persuade, rather than to impress, you see your own theories and experiments as your audience sees your theories and experiments. You become a critic of your own work. As Francis Bacon said, "Reading maketh a full man, conference a ready man, and writing an exact man."

So where do you get started? First, you need something to write about, either an idea for a proposal or results from your work. Although you don't have to wait until every result has been recorded, mapped, and plotted before you begin writing, you should not start writing a scientific or engineering document that has no destination.

The second thing that you need is an understanding of your constraints, particularly who your audience is and what the format will be. First, you must have an idea about who will read the document: what they know about the work, why they will read the document, and how they will read the document. Without an audience, you won't know what background information to include, which words to define, or how detailed to make your illustrations. You also should have an idea about the format. For instance, how long will the document be—two pages? twenty pages? two hundred pages?

Let's assume that you've got something to write about and that you understand your constraints. Let's say that you've spent the last fourteen months testing solid particles as a new heat transfer medium for solar power plants. The simplicity of the idea intrigues you. You envision a solar central receiver plant in a desert with hundreds of mirrors surrounding a central receiver. Falling through the receiver is a stream of silica sand that collects solar energy reflected from the mirrors and stores that energy in a tank below.

You've not only got visions; you've got results from simulation experiments in which concentrated sunlight has heated silica sand particles to more than 1000°C [Hruby, 1984]. You've also finished a computer model that shows the concept's thermal performance to be about 75 percent, a performance competitive with current central receiver designs [LaJeunesse, 1985].

The constraints of format and audience you also know. The audience is challenging because it is so varied: Department of Energy (DOE) officials, solar engineers, and utility engineers. Within this mix of audience, there are political considerations for the document. Over the past few years, funding for the solar energy program has dropped because current designs of solar central receivers have not proven efficient enough for utilities to invest in them. Management in the solar program is looking for some new ideas, and you see solid particles as a good one. Unfortunately, many officials at DOE want a quick-fix solution—they're antagonistic to something long-term such as solid particles. Nonetheless, you see solid particles as the best chance that solar energy has for competing with fossil fuels.

Your supervisor sees something too because she is eager for you to publish your work. "Have you written your report?" she asks you.

"No, not yet," you say.

Your supervisor stares at you for a moment. She has a casual look about her that's deceptive. Her office suddenly feels hot. "I'll be seeing something soon," she says.

"Soon," you say. "Soon."

You go back to your office where Hank Wilson is waiting. Hank Wilson wants to talk about the Braves. You remind yourself that during the last players' strike, you vowed never to watch another baseball game again. You remind yourself that your old heroes—Aaron, Cepeda, and Niekro—no longer play the game, and that a lot of the people who do play the game receive more money in a year than you'll earn in your lifetime.

Hank starts in about last night's game where the Braves came up with two runs in the bottom of the eighth to come back against the Reds. "It started with a couple of singles, both grounders with eyes, and then McGriff ripped a double to the gap."

Hank's story reels you in for a moment, but you remember your vows. The only problem is that the Braves have a three-game lead in their division, and as long as you can remember, you have been a diehard Braves fan. As a kid, the reason was silly: You wanted them to win the series so that you could go to a ticker-tape parade. As you grew older, you wanted the Braves to win a series for the simple reason that you had been waiting so long. Through the years of Aaron and Niekro and Murphy, you waited, reading boxscore after boxscore, staying up nights to catch radio broadcasts from the west coast, relishing each win, dying a little with each loss. You hesitate a moment, watch the light flicker in Hank's eyes, but don't give in.

Even if you wanted to talk about baseball, you can't. You've got results, you've got recommendations, and you've got a report to write. What should you do? You can't write here with Hank talking about baseball. The library, you think. "Catch me on it later, Hank," you say. "I've got to write my report."

You are almost out of the building when the department secretary stops you. "Al put something in your mail slot," she says.

"I don't want to look," you say.

"Then don't," she says.

For one instant, you think about moving on, but curiosity gets the better of you. Al is the department manager. In your mail slot is the program's *Quarterly Report*. It's your turn to review it. The *Quarterly Report* is the pride and joy of some paper pusher in the DOE. It contains almost nothing technical, only management things: milestone charts, procurement summaries, fiscal plots, and contract distributions.

"By tomorrow," she says.

"Tomorrow?"

"Don't look at me. That's what Al said."

"But I've got to write my report."

"That's fine," she says. "As long as this gets reviewed by tomorrow." The department secretary smiles at you. She has a wicked smile.

You go back to your office. Hank Wilson stares bewildered as you walk by. You try to ignore him. At your desk, you begin reading. You are only halfway through the standard DOE foreword when you catch yourself staring at the words and sentences, but not comprehending them. You sit up in your chair, go back to the first sentence, and begin again. Thirty minutes into your reading, a utility engineer from Arizona calls. His company is studying solar energy options, and he wants to know the thermal efficiencies of solid particle receivers. You promise to send him something tomorrow.

You push through the *Quarterly*'s milestone charts and fiscal sheets and begin checking the procurement summaries when an engineer from analytical modeling stops by. She wants to adapt your computer code to solve a problem in coal combustion. You force a smile and invite her in. She has several questions about your code—questions you try to answer off the top of your head, but can't. So you dig through your old notes and printouts.

By the time she leaves, it is five-thirty. You are tired and hungry. What now? You decide to go home and return here tonight, when no one is around and you can take the phone off the hook and write.

You get home and there's a bill from the power company. It's a bill you're sure you've already paid. You throw it in the wastebasket. Then you pick it out. There's something frightening about throwing away an unpaid bill. You decide to forget the bill and eat out. You need a good meal, you think. After all, you've got a lot of writing in front of you. You go to Pedro's for chili rellenos. While you eat, you read today's newspaper. You glance at the sports section. Glancing at the boxscores is okay, you decide. Reading about baseball is not actually watching baseball, and that was your vow: never to watch another baseball game. The Braves play in Houston this weekend. Houston has never been kind to the Braves. You know the reason, of course. It's the Astrodome. Baseball is supposed to be played outdoors.

It is seven o'clock by the time you finish eating and get back home. You're a little tired and you decide to rest a bit—just lie down on the couch and think about what you're going to write.

When you wake, it's quiet and dark outside. You look at the clock: 11:43. The day is lost.

Getting in the Mood

Scientific writing is a craft that requires preparation. It is difficult to jump right into writing a document. Most professional writers have disciplined schedules. Although you as a scientist or engineer cannot alter your entire lifestyle just to get one document written, you can make some simple adjustments.

The first thing you should do before writing is to

clear your mind. You have to put aside your personal problems and the bills you have due. You have to think about the work and the audience. One way to clear your mind is through sleep. Hemingway used this method. His writing day began each morning after eight hours of sleep. Another good way to clear your mind is through motion: walking, running, mowing the lawn. Wright Morris walks twice a day, before and after writing. Harper Lee plays golf. Max Apple and John Irving are avid runners. Mowing is okay, but you've got only so much lawn. Driving lets you daydream, but driving and daydreaming at the same time is dangerous. Running, walking, and bicycling are the best. They give you the chance to think about the structure of the document, play a strategy through in your mind, and see whether it makes sense. I prefer a steady run, about thirty to forty minutes, not so long that I'm exhausted, but long enough that I begin daydreaming. After taking a run or walk, I don't eat much more than a snack. Then I shower and slip into something casual.

The second thing you need is a block of free time. For drafting a document, a good writing block is one to four hours. Even after sitting down at your desk, you usually need ten minutes or so just to get in the groove of actually writing sentences. Don't think, though, that you can continue writing for the whole day. After a while, your concentration fades, and your writing efficiency diminishes.

Finding a block of time at work is often difficult. There are so many interruptions: phone calls, e-mail messages, visitors. To make the writing process efficient for yourself, you should eliminate those interruptions. You should shut the door, unplug the phone, mute the electronic mail messages, and concentrate on getting your thoughts onto paper. You will find that absolute silence in your room is not necessary. White noise is okay. So is soft instrumental music. However, each interruption, no matter how short, will cost you an additional five or ten minutes of writing time because it takes you that much time to get back into the document—to find your place again and begin writing sentences that add to what you've already done.

Some distractions are created, rather than imposed. A few years ago, computer games, especially the game of bridge, posed a problem for me. If that game was on the computer, I felt a constant force tugging me away from my writing. Just a window away was a freshly dealt hand. After struggling for a few months to achieve a balance between work and play, I went cold turkey one day and deposited all my computer games in the trash. The result was that my writing productivity increased dramatically.

Finally, you have to prepare yourself mentally for the task ahead. Scientific writing is hard work. It is not something that you can casually begin after a heavy meal or a couple of drinks. Many scientists and engineers mistakenly think of professional writers as free spirits who write perhaps two days a week and play the other five. In this imagined schedule, writers spend most of their time fishing, traveling, or going to bull fights. Actually, most professional writers are diligent workers. Hemingway, for instance, followed a schedule of writing for eight hours a day, six days a week.

Instead of worrying about the way that professional writers spend their leisure time, you should consider the way that professional writers spend their writing time. For instance, you should look at how professional writers structure their sentences and paragraphs. Don't, however, choose writers from the eighteenth century whose language is outdated, or writers such as Faulkner or Joyce whose styles are luxuriant. Rather, choose writers whose styles are crisp and straightforward. Richard Rhodes, who authored *The Making of the Atomic Bomb*, is an excellent choice. So is Annie Dillard, author of *Teaching a Stone to Talk*, or David Simon, author of *Homicide*.

Writing First Drafts

The next day, you quickly take care of the small things at work: your critique of the *Quarterly Report*, a memo to the utility engineer in Arizona. You leave work early and eat a light meal at home. You wash the dishes, put in a load of laundry, and go out for a slow run—something to clear your mind. You run out beyond the subdivision, out into cattle land where you're surrounded by barbed wire and brown grasses. There is still daylight, and you are thinking about what you are going to write.

After you finish your run, you shower and drive back to the office. No one is there. It is almost seven o'clock. Somehow the place seems different. Maybe it's the stillness, or just the quiet. The only sound is the hum of the drinking fountain. You sit at your desk with your spiral notebook of results in front of you. You are in the mood.

You turn on the computer and stare at the blank screen. You finger the keyboard, but do not type anything. You are still thinking. The steady, expectant hum of the computer bothers you so much that you turn toward your other table and grab a pencil. You need something written down before you can start on the computer. You write a title for your paper:

Solid Particle Solar Receiver Design

Tiresome, you think. Too many big words in a row. You erase and try again.

Solid Particle Receiver Design

That's too confusing. Most people wouldn't know that you are talking about solar energy. More important, the research isn't really the design of a solid particle receiver, but whether solid particles could work as the heat transfer medium in a solar central receiver. You write another title:

Using Solid Particles in Solar Central Receiver Systems

It's not great, but it will get you started. You put down your pencil and pick up your pen. Ink looks so much more professional than lead. You rewrite the title in ink and then write your name below the title. The rest of the page is blank. It stares back at you. Suddenly, you feel tired. You lean back in your chair and rub your eyes. You loathe first drafts. You would do almost anything instead of writing a first draft: fill out your taxes, clean the bathtub, watch a stock car race on television.

Come on, you tell yourself. You've wasted enough time. You must start. You must write something down. An outline, you think. That's what you were always taught. So you write one:

A. Introduction
B. Experiment
C. Computations
D. Conclusions

What about a summary. You write "Summary" above "Introduction." What letter should you assign it? You already used A for "Introduction." You decide to make it Z. You're not happy with Z, but you see little choice besides crossing out the other four. You wish that you had stayed with the pencil. Now you look at your outline.

Z. Summary
A. Introduction
B. Experiment
C. Computations
D. Conclusions

Not much help you think.

You decide to try something else. You write "Summary"at the top of a blank page, then "Introduction" at the top of another page, and then "Experiment," "Computations," and "Conclusions" at the top of three more. You pick up your five outline pages as well as your title page, bunch them into a stack, and then place them back down. Now you're getting somewhere.

You put aside the title page and start with the summary page, but you remember that an informative summary is the last thing that you write. So you put it aside and look at your introduction. You start writing a sentence about how the solar central receiver program began in 1976, but then you stop. What's an introduction supposed to do? Introductions are supposed to prepare readers for what's ahead. How can you prepare someone for something ahead if you don't know what that something is. You put the "Introduction" page aside as well.

So far, things aren't going well.

For some reason, you are restless. It is eight o'clock and you are hungry again. You go down the hall and get a candy bar. You know that candy bars are bad for you, but you're working hard tonight and you deserve a treat. After you finish the candy bar, you feel tired. You consider calling Hank Wilson to casually ask whether he watered the office plants today, although you know he hasn't. You have the hope that he might volunteer the Braves score. Then you think about going home and crawling into bed and sleeping until you can't sleep anymore.

You have lost the mood.

Don't deceive yourself. The first draft of a document is difficult. In the first draft, you juggle your topic and your writing constraints with all three elements of style. What should you do to get through a first draft? The first thing you should do is take a few deep breaths and prepare yourself for the road ahead. If the document is long, you're not going to finish the draft in one sitting. If you do, chances are that the draft won't be any good. In writing, you have to be realistic. It has taken you several months to do your work It's also going to take some time to finish the document—days in some cases, weeks in others.

Now for the actual act of writing the first draft. Should you use a pencil, pen, or computer? As John Gardener [1983] said, that question is much deeper than it sounds. The real question, he says, is how do you get words onto paper. Regarding the surface question of which writing instrument to use, you should use whatever it takes. A computer certainly has the most advantages. On a computer, it's easy to add text, delete text, bring in text from other documents, and incorporate illustrations. Still, some people feel more comfortable beginning on sheets of paper, at least for the brainstorming stages. Writing on sheets of paper has the advantage of allowing you to see the document's entire strategy on your desk without having to scroll through the document on the computer. Also, many people feel uncomfortable sitting before a computer and just thinking, which working on the initial stages of a document requires you to do.

What about the deeper question of how you get words onto paper? The answer is the same: whatever it takes, short of plagiarism. Should you use an outline? Every English teacher you've ever had has probably told you to use one. That fact alone should make you suspicious. The answer to the question is a qualified "yes." A strong outline is helpful for long and complicated documents. A strong outline gives you a structure before you begin writing. That way, in your early drafts, you have to juggle only two elements of style—language and illustration—rather than three. To be effective, though, your outline must be strong. What makes for a strong outline? Detail. A strong outline not only lists your section headings, but also your subsection headings, notes about each section, and some sentences or phrases you might use. Outlines by professional writers usually don't look like the neat little ABCD things your English teachers put on the blackboard. Professional outlines, such as the one shown below, are long and spread out and above all, filled with the kinds of things that make it easy for you to write a first draft.

OUTLINE

USING SOLID PARTICLE RECEIVERS
IN SOLAR CENTRAL RECEIVER DESIGNS

Audience

The principal audience is DOE headquarters. They have a general knowledge about solar central receivers, but not about solid particle research. Some DOE members are antagonistic to the solid particle idea (they want a quicker solution).

The secondary audience includes solar engineers at the utilities and national laboratories. These people are knowledgeable about solar central receivers and mainly want to know whether the solid particle idea can work.

Introduction

Identity of research:

This report will discuss using solid particles as the heat transfer medium in a solar central receiver. Requirements for such a heat transfer medium include

(1) Heat transfer medium must withstand temperatures up to 800°C
(2) Convective losses from medium must not be too high (define "too high")
(3) What else?

Reasons that research is important:

(1) Diversifies solar energy applications by obtaining higher temperatures than those obtainable from current designs
(2) Allows for high-temperature experiments and modeling

Background material (heat transfer media):

(1) *Water/steam*. Advantages: utilities familiar with; test data available. Disadvantages: not efficient enough; difficult to control in two-phase flow.
(2) *Molten Salt*. Advantages: relatively high heat capacity; high storage capacity; single-phase fluid. Disadvantages: freezing in pipes; corrosive nature.
(3) *Liquid Sodium*. Advantages: high heat transfer capacity. Disadvantages: low storage capacity; safety problem.
(4) *Solid Particles*. Advantages: very high heat transfer capacity (define "very high"); absorbs radiation directly; transfers heat directly. Disadvantages: abrasive nature; utilities less familiar with.

Mapping of Discussion

Discussion

.
.
.

Note that this outline has parts of sentences interspersed among the details. If you are writing an outline and suddenly come up with an idea for a sentence or transitional phrase, it makes sense to include that sentence or phrase in your outline. If not, you risk the chance of forgetting the sentence or phrase when you begin the first draft. What if someone else asks to see the outline, though? Won't all of these details and questions and sentence parts confuse them? Probably. Sometimes supervisors ask to see the outline of a document to check the progress of that document. In such cases, you should clean up your outline so that your supervisor can see the strategy. As it turns out, your "cleaned up" outline will often look like the ABCD outline that your English teacher taught you. Still, your "working" outline should be in a form that helps you write the first draft. If your working outline includes green ink for illustration ideas and red ink for ongoing questions, so be it.

How do you actually move from the outline to the first draft? Let's first assume that you have a strong outline and that you are in the mood to write. Now, there are two basic approaches to writing the first draft: the rabbit's and the turtle's.

Rabbits hate first drafts. They despise juggling the constraints of writing with all the elements of style. So, in a first draft, they sprint. They write down everything and anything. Rabbits strap themselves in front of their computers and finish their drafts as quickly as possible. Unfortunately, their first drafts are horrendous, sometimes not much better than their outlines. Still, they've got something. They've put their ideas into a document, and they're in a position to revise.

Turtles, on the other hand, are patient. Turtles accept the job before them and proceed methodically. A turtle won't write down a sentence unless it's perfect. In the first sitting, a turtle begins with one sentence and slowly builds on that sentence with another, and then another. In the second sitting, a turtle then goes back to the beginning and revises everything from the first sitting before adding on. It usually takes a turtle several sittings to finish a first draft, but the beginning and middle are smooth because they've been reworked so many times. Revision then entails looking at the document from an overall perspective and smoothing the ending.

Which type of writer should you be? The answer is whatever type works for you. Actually, few professional writers are strictly rabbits or turtles. Most professional writers are a combination. If you become stuck in the first draft and your writing slows to a stop, you should become a rabbit so that you can finish. On the other hand, if you're working on a section that establishes the organization for the rest of the paper, you should be a turtle; otherwise, you might have to start over.

Which section of a paper should you begin writing? The introduction or the discussion? This question is difficult. On the one hand, you psychologically need some kind of introduction before you can begin the discussion. On the other hand, the discussion is the easiest section to write first. My feeling is that you should be a rabbit on the first draft of your introduction and then write the discussion and conclusion at your normal pace. After finishing the conclusion, you should go back and rewrite the introduction. What about your summary? A descriptive summary you can write first, but an informative summary you want to write last. One good way to draft the informative summary is to wait until you have a polished draft of the text. At that point, you can highlight the most important sentences and illustrations of the draft. With a little transition and background added, these highlighted sentences and illustrations can serve as a first draft for the informative summary.

The key to writing first drafts is momentum. Because

scientific writing is so textured (in other words, because each sentence depends so much on the ones around it), you need to sustain momentum during your first draft. Otherwise, you'll lose your train of thought. How do you sustain momentum? Well, here are a few habits that you can borrow from professional writers:

First, set realistic goals. How large a goal depends on how much of a rabbit or turtle you are. If you're a rabbit, then a realistic goal for a single sitting might be five pages. If you're a turtle, a realistic goal might be one page. Realistic goals are important. Psychologically, you want to end each sitting with a feeling of accomplishment.

Second, end your sittings by writing into the next section. That way you'll find it easier to start writing the next time you sit down. You will have a little momentum rather than a blank page. The fiction writer André Dubus claims that when he ends a sitting, it is mid-sentence, mid-thought.

Third, watch what you eat. Writing makes you restless, which makes you hungry. One way to counteract that is to sip on a tall glass of water while you write. If you must eat, don't eat foods that will make you sleepy. Also avoid salty foods that make you thirsty or foods that require two hands such as bananas (you have to peel them). You want something that you can reach with one hand. Celery and carrot sticks are probably the best, but you can munch on them for only so long. Raisins aren't too bad; neither are unsalted sunflower seeds. Since the advent of breadmakers, I've been opting for slices of fresh bread.

Finally, when you finish a draft, store it on your computer under a separate file name. Having a clean version that you can print out at any time gives you a psychological advantage when you start revising the original. You won't worry about being caught empty-handed in case your supervisor needs a clean draft right away.

How do you write a first draft when you're pressed for time? There are really no shortcuts in writing a first draft outside of abandoning your friends and family and locking yourself in a room until it's done. Actually, when you're pressed for time, the one part of the writing process that you don't want to cut short is the first draft. If anything, you should spend extra time on your first draft because you can't afford to choose a structure that is inappropriate. Choosing an inappropriate structure could cause you to have the document in pieces on your desk when the deadline strikes.

What do you do if you've got writer's block? Writer's block is one of those bits of folklore that many scientists and engineers talk about, especially when they're having problems writing. "The words just aren't flowing," they'll say. "I must have writer's block." Well, the words rarely "flow" for anyone, professional writers included. Most professional writers struggle the same way you do. They struggle with paragraphs and sentences and words. So what is the source of this term "writer's block"? When writing a document, there are certain things, besides distractions, that cause you to stop.

First, you can't think of the right word. Not having the right word happens to everyone. Many times it's on the tip of your tongue. In such a case, a possible solution is to pull out a dictionary and spend two or three minutes looking. If you don't find the word after three minutes, write down the name of your favorite baseball player (mine is Roberto Clemente) and keep going. Your subconscious will think of the right word later.

Second, you can't find the sentences to express an idea. In this situation, chances are that you haven't quite grasped that idea yet. A possible solution is to skip two lines and keep writing. Let the idea simmer. Maybe you can think about it on your next run or walk or exercise activity. Besides getting you in the mood to write, exercise is an excellent activity for helping you sharpen your ideas.

Third, you hear voices. When many scientists and

engineers sit down to write, they hear voices, critical voices: their eighth grade English teacher, their department manager, the theorist from Oxford who complains bitterly about everyone else's writing but doesn't write so well himself. These voices inhibit you—you just can't make your pen write another word. A possible solution is to put on your favorite classical music and turn up the volume until it drowns the voices. Bach's *Brandenburg Concertos* and Mozart's piano sonatas work well for me.

In scientific writing, there really is no such thing as writer's block. Once you have done the work, the ideas are there. Writing those ideas is more mechanical than inspirational. For scientists and engineers, the real block occurs when the ideas won't come, but that block is a block in science and engineering, not in writing.

Revising, Revising, Revising

Revision is the key to strong scientific writing. Revision is the difference between a title such as

Using Solid Particle Receivers
in Solar Central Receiver Systems

and a title such as

Using Solid Particles as the Heat Transfer Medium
in Solar Central Receivers.

For the work described in this chapter, the second title is more precise. You weren't deciding whether solid particle receivers could work in a solar central receiver system. Rather, you were deciding whether solid particles could work in a solar central receiver. This difference is important, but this difference might not have been apparent when you were staring at the blank computer screen of a first draft.

In your first draft, you have to juggle too many elements of style to give proper attention to any one. Revision gives you the luxury of considering on any single draft only the parallel structure of your headings or the conciseness of your language or the clarity of your illustrations. Scientists and engineers, perhaps more than other professionals, understand the importance of revision. In their experiments, scientists and engineers depend on trial and error. In their computations, scientists and engineers use many iterative techniques to arrive at solutions. Just as those techniques are important to science and engineering, so is revision important to scientific writing.

Many scientists and engineers hold the misconception that great writers don't have to revise. These scientists and engineers assume that great writers think only great thoughts and then effortlessly write them down in a prose that needs no reworking. Well, maybe there have been some great writers such as D. H. Lawrence whose first drafts could go straight to press, but Lawrence and writers like him are in the small minority. For every D. H. Lawrence, you can find ten writers who were constant revisers: Leo Tolstoy, Ernest Hemingway, Anton Chekhov, Ralph Ellison, S. J. Perelman, Carolyn Chute, E. B. White, Flannery O'Connor, Raymond Carver, Evan Connell. Raymond Carver claimed that he never took fewer than ten drafts to finish a story. If great writers such as Raymond Carver required ten drafts to smooth their writing, just think how many revisions the rest of us need. In my own writing, I average about five pages a day. Unfortunately, they're all the same page. It's unusual for me to have a sentence that was written on the first draft make it to the final draft unchanged. For most of us, the key to successful writing lies in working hard on our revisions, not in conjuring magic on our first drafts.

What does revision entail? Before you begin revising a document, you should obtain some distance from it. After finishing a first draft, spend a day working on another project. Better still, hike in the mountains or have a night on the town. You need a few hours, sometimes a few days, away from your first draft so that you can effectively revise. Things have to settle. During this time, ideas for changes will probably come to you. Write these ideas down, but don't go back to your draft.

Let's assume that you have some distance between you and your draft. What next? An important thing for you to do is to change your personality. You should not have the same personality when you revise as you did when you wrote the first draft. Whether you worked as a turtle or a rabbit on the first draft, you were erecting the document. In the revision stage, you have to chisel and polish what's on the page.

To become a successful reviser, you have to become a good reader. How do you become a good reader? One way is to listen to other good readers. Just as no two people write the same way, no two people read the same way. Some people are particularly attuned to language. Others are sensitive to structure. Listening to good readers review a document will help you strengthen your own reading. A second way to become a good reader is to practice. Work hard on your critiques of other people's documents. Be specific. Don't just tell these people that something is unclear. Show them why you became confused. Also, don't just mark weaknesses in a document; mark the aspects that you find successful. You will find that telling someone what is strong in a document is as difficult as pointing out the things that are weak. How quickly you mature as a writer will depend on how quickly you mature as a reader.

What's the best method of revising? There is no definitive answer. Still, there are some techniques that many successful writers adopt:

One thing to do is to change the look of your document when you revise. If you drafted your document on a computer, try revising on a hard copy. If you drafted your document in your office, try revising the document in the library or conference room. In other words, make the revision process different from the drafting process.

Second, try to work through large chunks in each sitting. Revising large sections of your document in single sittings makes your document smooth. You'll see gaps in logic and faults in structure. During your first few revisions, you'll be writing a lot—filling in spots. As with the first draft, set realistic goals for yourself on each sitting. In later revisions, you should have fewer and fewer corrections, and therefore you can read larger chunks. Ideally, you'd like to reach the point where you can read through the entire document in a single sitting or a single day.

Third, get some distance between each revision. Avoid making multiple passes through a large document on the same day. If you read a document or section over and over, it's easy to go stir-crazy and start changing things for no reason. How many times should you revise a document? For your documents to really sing, you should revise until you are nitpicking over language: a word here, a comma there. In your later drafts, try reading the most important sections aloud. You'll be surprised at how many things you'll catch with your ear. Although you want to get some distance between drafts, you don't want to get too much distance. Allowing a draft to sit on a desk for months certainly gives you distance from the draft, but also saps the momentum from the project. You'll feel compelled to update the science and engineering in the document, which will make the revision seem more like a first draft.

Finally, solicit criticism of your writing. Soliciting criticism is an excellent way to revise. No matter how you try to separate yourself from your work, you will take certain things for granted. For instance, in your report about solid particles, maybe you forgot to mention the composition of your solid particles in the summary (your solid particles were silica sand). That detail would be an easy oversight for someone engrossed in the work, but one that a fresh reader would quickly catch. If you are going to solicit criticism, you should be willing to make changes. This caution might sound silly, but you wouldn't believe how many times someone has asked me to review a document and then been appalled when I actually suggested revisions.

Once you solicit criticism of your writing, you should be prepared to accept that criticism. Don't be defensive when people whom you respect speak harshly about your writing. They're not attacking you. Actually, when people become harsh over your writing, it's a compliment of sorts. It means they're frustrated—they're genuinely interested in your work and they want your writing to inform more efficiently. If your critics are indifferent, that's the occasion to be upset.

How do you know when to incorporate criticism and when to dismiss it? This question is difficult. You shouldn't give a rubber stamp to every suggested change, even if the suggestions come from a good critiquer. Rather, you should weigh all criticisms. If you think a criticism is valid, incorporate it. If you think a criticism is "off the wall," at least mull it over. Maybe there was a deeper problem within that section or paragraph that the critic couldn't articulate.

Don't seek too many critics. You'll get confused. On a five-year plan for one of its solar programs, the Department of Energy decided that everyone in the program should review the document. In effect, the Department of Energy wanted all the readers to become the writers. The result was chaos. Because the report was general in its original scope, the two hundred critiques only served to push it into two hundred different directions. For several years, the Department of Energy haggled over draft after draft of that report, wasting thousands of man-hours, before finally scrapping the whole document.

This example raises the question of how you should approach collaboration on a document. Collaboration has both advantages and disadvantages. The principal advantage is that working in a group broadens the range of ideas that the document can incorporate. Collaborative writing also allows the group to draw from the various editing strengths of the members, such as one person's ability to see the big picture or another's ability to smooth transitions between ideas or yet another's ability as a proofreader. The principal disadvantage of collaborative writing is an inherent lack of continuity. With multiple writers contributing to the text, you can have inconsistencies in language, particularly in word choices, average sentence lengths, and degrees of sentence variety. In an ideal collaboration, you optimize the advantages and minimize the disadvantages.

One way to accomplish that goal is to solicit everyone's ideas and strategies at the outlining stage, but then have one person be the lead writer for the project to insure that the document maintains continuity. In the revision stage, the group would again have input. For this collaborative scheme to work, the group should respect the writing ability of the lead writer, and the lead writer should respect the editing comments of the group. This scheme is not the only way to achieve a successful collaboration. More important perhaps than the collaboration scheme is the interaction of personalities. Each person in the group should realize that the final product will not read exactly as he or she would desire. In other words, everyone in the group has to bend a little. That flexibility does not mean that group members should tolerate illogical strategies or discontinuous language, but that group members should accept that there are multiple ways to organize a section and achieve sentence variety.

What happens to revision if you're on a tight schedule? In many projects, the time allotted for the writing is just too short. You might have begun with good intentions (for instance, scheduling a month to write the report), but then you lost twelve days in February because of a late shipment, five days in April because of a sick technician, three days in May because of a vacuum leak, and another three days in July because the computers were down. The month that you scheduled for writing has slipped to one week. So what do you do? Well, as stated in the previous section, you should spend enough time on your first draft to secure a structure that works. Then you've got to revise efficiently.

Revising a document is much the same as leveling a piece of land. In your first couple of revisions, you wield a pick for groundbreaking as you edit on the section and subsection level. Then in your remaining revisions, you use a rake for smoothing, as you progress from rearranging paragraphs to polishing sentences to nit-picking over words. When time is short, you can't afford to revamp the structure in your last couple of revisions. If you do, you might improve the overall grade a little, but you'll unsettle the soil a lot. Also, when pressed for time, you should reduce the time between revisions. In this reduction, you should find some absorbing activity such as running or walking or perhaps a light meal away from the office to separate yourself from the previous draft.

At some point in the process, you have to say "enough"—enough to outside suggestions, enough to revisions. You have to decide that your document is successful and then you have to finish it.

Many scientists and engineers think that perfection can be achieved in writing. Except perhaps for the ten commandments, there is no perfection in writing (and even the ten commandments suffered revision between Exodus and Deuteronomy). Words are not exact substitutes for thoughts. There will inevitably be something that you will want to change: some sentence that won't sound right on Wednesday, another sentence that won't sound right on Thursday.

Although there is no perfection in writing, there is success. For your sanity, you have to find a point at which you stop drafting. Evan Connell once said that he knew he was finished with a short story when he found himself going through it and taking out commas, and then going through it again and putting commas back in the same places. The documentation of strong science and engineering deserves that amount of care. Although you can't achieve perfection in your writing, you should still strive for it. Strive for perfection, but be content with success. One of the beauties of writing is that you don't stop learning. With each document, you improve your craft.

Finishing

It is two o'clock in the morning. Your report has gone through the laboratory's sign-off and you have finished proofing it for what seems like the hundredth time. You are exhausted, but you correct the last two typos and print out the final copy. No one is in the office. You wish someone was—anyone. They wouldn't have to read your report, just look at it, admire how clean and neat it is. You'd tell them how smooth the writing was, maybe read aloud the title and summary.

Finishing a large document is taxing work. By your final draft, the words seem dead on the page. You don't really read them. You only see yourself reading them. Although the final proof of a manuscript is often tedious, you can't let up. A small mistake such as a misspelled word will unsettle your readers and undercut the authority of your work. Finishing a paper is much the same as finishing a baseball game. Some teams, when they're ahead, let up during the last few innings. They play sloppily, sometimes so sloppily that they lose their lead. Some writers are the same way. They work hard on the first few drafts, and then let up on the final drafts, allowing typos to pull down their work.

Should you rely on a computer spell-checker to find those typos? "Rely" is the wrong word here. Spell-checkers are wonderful for catching some misspellings such as "recieve" instead of "receive." Some spell-checkers even notice if you repeat a word such as "in the the chamber." However, spell-checkers miss a number of misspellings, such as when the misspelling is another word ("to" rather than "too"). Spell-checkers also do not catch errors such as missing words ("in chamber" rather than "in the chamber"). For these kinds of errors, you need a person rather than a computer. In other words, you should use a spell-checker, but that spell-checker should supplement, rather than replace, your own careful proofreading.

Besides being hard work, the completion of a large document is often thankless work. By the time your document appears in print and people comment on it, the feelings that you have for the writing have usually been lost. You are probably absorbed in your next project. That's the way the writing game is. Don't expect satisfaction in your writing to come from other people. You'll be sorely disappointed. Satisfaction in your writing has to come from within. It is the realization that you've done good work.

You make a final copy of your report and place it in the mail for the print shop. You are euphoric and exhausted at the same time. It is over. You leave the lab. The night is cool, and the moon glimmers behind some thin clouds. You think you hear opera. You get in your car—a maroon and white Monte Carlo with a bad muffler—and drive past the laboratories, past the open fields tinted purple and blue by the night. At a stop sign, your muffler wakes a neighborhood dog. You wish that you could wake everyone and tell them what you've done.

You're tired, but you can't go to bed—not just yet. You need some kind of celebration. Pedro's, you think. You'll order a chili relleno and read the sports pages. The Braves are headed for the playoffs, and this time you've got a good feeling that they're going to win.

References

Dillard, Annie, *Teaching a Stone to Talk* (New York: Harper-Collins Publishers, 1982).

Gardner, John, *On Becoming a Novelist* (New York: Harper and Row Publishers, Inc., 1983), p. 119.

Helmholtz, Hermann, in "Ziele und Wege des Sprachschatzes im ärztlichen und naturwissenschaftlichen Schriftum" by Erwin Liek, *Münchener medizinische Wochenschrift*, vol. 69 (1922) p. 1760. Original: *"Die schriftliche Ausarbeitung wissenschaftlicher Untersuchung ist meist ein mühsames Werk...Ich habe viele Teile meiner Abhandlungen vier- bis sechsmal umgeschrieben, die Anordnung des Ganzen hin- und hergeworfen, ehe ich einigermaßen zufrieden war."*

Hruby, J. H., and B. R. Steele, "Examination of Solid Particle Central Receiver: Radiant Heat Experiment," *Proceedings of the ASME-ASES Solar Energy Conference* (Knoxville, TN: March 1985).

LaJeunese, C. A., *Thermal Performance and Design of a Solid Particle Cavity Receiver*, SAND85-8206 (Livermore, CA: Sandia National Laboratories, 1985), p. 56.

Rhodes, Richard, *The Making of the Atomic Bomb* (New York: Simon and Schuster, 1986).

Simon, David, *Homicide* (New York: Fawcett Columbine, 1991).

Appendices

Appendix A

Avoiding the Pitfalls of Grammar and Punctuation

For a long time, I had difficulty making good grades in English classes because I couldn't reconcile the rules my English teachers would teach in composition classes with the abuses of those rules that I found in literature classes. Then one day I realized that just as there are different levels of dress that depend on the formality of the occasion, so too are there different levels of grammar and punctuation rules that depend on the formality of the writing. For that reason, Mark Twain could use fragments and contractions in *The Adventures of Huckleberry Finn,* but I couldn't use those same fragments and contractions in my English compositions.

Many mechanical rules, such as the rules governing pronoun references, are intended to eliminate ambiguities. Other mechanical rules, such as using a plural verb with a plural subject, meet readers' expectations. However, some grammar rules, such as not ending a sentence with a preposition or not splitting infinitives, are more difficult to justify. In some cases, eliminating the preposition from the end of a sentence improves the sentence by tightening it. In other cases, though, not having a preposition at the end convolutes the sentence, as in Winston Churchill's sarcastic remark about the rule: "Ending a sentence with a preposition is something up with which

I will not put." In this second class of cases, the rule becomes akin to wearing a tuxedo when a simple coat and tie is sufficient. Nonetheless, many managers and editors expect us to follow these tuxedo rules of grammar and punctuation even though most readers no longer have them as expectations. As a writer, you have two choices: 1) break the rule and say "to hell with" what these tuxedo critics think, or 2) work around the rule to appease the tuxedo critics.

Although tuxedo rules cause a lot of consternation between editors and writers, tuxedo rules themselves are noticed by only a few readers. Other grammar and punctuation rules, though, are noticed by most readers and are needed for efficient writing. This appendix presents a review of rules from the latter category. What about using the computer's checker for grammar and punctuation? As with the computer's spell-checker, the checker for grammar and punctuation is useful, but cannot do the job alone. Such a checker can find certain mistakes such as subject-verb disagreements, but not others such as dangling modifiers. In the end, you must decide whether the sentence is correct.

Avoiding the Common Pitfalls of Grammar

Grammar is the system of rules by which words are arranged into meaningful sentences. Although many English-speaking people complain about grammar, the number of grammatical rules in English is small compared with the number of rules in other languages such as German. The following rules address common pitfalls of grammar in scientific writing:

1. *Do not join two independent clauses with an adverb.* Sentences are the fundamental units of expression in scien-

tific documents. Readers expect you to write in sentences. When your sentences run on, your readers lose their place in the paragraph. They also lose their confidence in you.

The most common type of "run-on" occurs when the writer tries to use an adverb, such as "therefore" or "however," to join two independent clauses. In such cases, you should do one of the following: (1) begin a second sentence, (2) join the clauses with a semicolon, or (3) join the clauses with a coordinating conjunction such as "and," "or," or "but."

Mistake:	There is no cure for Alzheimer's, *however,* scientists have isolated the gene that causes it.
Correction:	There is no cure for Alzheimer's. However, scientists have isolated the gene that causes it.
Correction:	There is no cure for Alzheimer's; however, scientists have isolated the gene that causes it.
Correction:	There is no cure for Alzheimer's, but scientists have isolated the gene that causes it.

2. *In a list, present the items in a parallel fashion.* As with the first rule, this rule is important because of reader expectations. If your first slice of pie is apple, then readers expect the remaining slices to be apple.

Mistake:	The process involves three main steps: cooling, chopping, and *pulverization.*
Correction:	The process involves three main steps: cooling, chopping, and pulverizing.

3. *Have modifiers point to the words that they modify.* Failure to follow this rule causes ambiguities.

Mistake:	First, you find a latent print. *After being detected,* you dust with the powder.
Correction:	First, you find a latent print. After detecting the latent print, you dust with the powder.

4. *Have each subject agree in number with the verb.* When you have a singular subject, readers expect you to use a singular verb, and when you have a plural subject, readers expect you to use a plural verb.

Example: A series of shocks often precedes a large earthquake. *(Singular subject, singular verb.)*

Example: Two aftershocks of the earthquake were almost as powerful as the earthquake itself. *(Plural subject, plural verb.)*

Deciding whether some subjects are singular or plural is not straightforward. For instance, compound subjects are sometimes treated as single units:

> Under these conditions, the simultaneous seeding of the fluid's flow and measurement of the fluid's temperature *is* difficult.

Also, some foreign words such as "criterion" (Greek), "phenomenon" (Greek), and "stratum" (Latin) have unusual plurals: "criteria," "phenomena," and "strata."

Mistake: The phenomena *was* studied.
Correction: The phenomena were studied.

Moreover, words such as "none," "some," and "all" are singular in some instances, but plural in others.

Example: Some of the water was wasted.
Example: Some of the dolphins were infected.

Finally, if the subject consists of two singular nouns joined by *or*, *either...or*, or *neither...nor*, the subject is singular and requires a singular verb.

Example: Neither oxygen nor nitrogen is a noble gas.

If the subject consists of two plural nouns joined by *or*, *either...or*, or *neither...nor*, the subject is plural and requires a plural verb.

Example: Neither ceramics nor gases conduct electricity at low voltages.

If the subject consists of a singular noun and a plural noun joined by *or*, *either...or*, or *neither...nor*, the number of the second noun determines whether the verb is singular or plural.

Example: Neither the pilot nor the crew members were present. *("Crew members" is plural; therefore, the verb is plural.)*

5. *In each section of a document, maintain the same reference frame for the tenses of verbs.* If in a section, you assume that an event of that section occurred in the past, then that event should remain in the past for the entire section:

> ***Experiment.*** The experiment consisted of a Wolfhard-Parker burner in a stainless-steel container. The burner slot for the fuel flow was rectangular and was surrounded on all sides by passages for flow of air. Previous experiments had shown that such a geometry provides a nearly two-dimensional flame.

Because the first sentence of this section places the experiment in the past tense, all details in this section occurring during the experiment are in the past tense, and all details occurring before the time of the experiment are in the pluperfect tense (for example, "had shown"). Notice that the last detail ("provides") is in the present tense because it is a time-independent fact.

Should the writer choose to have the experiment occur in the present tense, the reference frame shifts up one notch, as do all verbs except those presenting time-independent facts.

> ***Experiment.*** The experiment consists of a Wolfhard-Parker burner in a stainless-steel container. The burner slot for the fuel flow is rectangular and is surrounded on all sides by passages for flow of air. Previous experiments have shown that such a geometry provides a nearly two-dimensional flame.

Avoiding the Pitfalls of Punctuation

Punctuation rules are important. They have been devised to eliminate ambiguities in language. Pay attention to the way strong writers such as William Safire use punctuation marks. Few things undercut the authority of a piece of writing more than a simple mistake in punctuation.

The Period. The period is the most powerful piece of punctuation at your disposal. In many scientific documents, periods are not used often enough. Too many sentences go on and on, taxing the reader's concentration:

> For temperatures above 1100K, the four fuels examined had about the same ignition delay when the ignition delay was defined as the time to recover the pressure loss from fuel evaporation, in spite of the large variations in ignition delay among the four fuels at lower temperatures.

There are too many ideas packed into one sentence. Clarity demands more than one sentence.

> Ignition delay is the time required to recover the pressure loss from fuel evaporation. Despite the large variations in ignition delay at lower temperatures, the four fuels had about the same ignition delay for temperatures above 1100K.

Although you should generously use periods to apportion your ideas into separate sentences, you should avoid using periods to abbreviate. When used in abbreviations, periods often trip readers; readers think they've come to the end of the sentence:

> Fig. 1.1 shows a gamma-ray line, i.e., radiation at a single gamma-ray energy level, that theorists had predicted would result from N. Cygni.

This sentence is choppy. By varying punctuation and cutting needless abbreviation, you can make a much smoother sentence.

> Figure 1-1 shows a gamma-ray line (radiation at a single gamma-ray energy level) that theorists had predicted would result from Nova Cyngi.

The Comma. Commas cause headaches for many scientists and engineers. Some scientists and engineers paint their sentences with commas. These scientists and engineers use commas anywhere there's the slightest suggestion of a pause. The result is that readers must wade through each sentence.

> Although many warnings, from governments, have been issued about acquired immunodeficiency syndrome, also knows as AIDS, we think that, for, at least, the next decade, its incidence will continue to increase.

This sentence reads too slowly. You *must* cut the commas after "warnings," "governments," and "that." You *could* also cut the commas surrounding "at least" and following "decade."

Other scientists and engineers scorn commas. These scientists and engineers will use commas only in the most extreme cases, and sometimes not even then. The result is that readers trip over ambiguities.

> After cooling the exhaust gases continue to expand until the density which was high in the beginning reaches that of freestream.

This sentence needs a comma after "cooling" and a set of commas around the clause "which was high in the beginning." For more discussion about using commas with the word "which," see Appendix B.

There are many rules for commas. Some rules are straightforward. For instance, you should use commas to set off contrasted elements (these expressions often begin with "but" or "not").

> The shark repellent with 20% copper acetate and 80% nigrosene dye was quite effective against Atlantic sharks, but ineffective against Pacific sharks.
>
> Many injuries result from shark bumps, not shark bites.

Also, in a series of three or more items you should use commas to separate each item. Therefore, write

> ...neopentane, perdeuteroneopentane, or neooctane.
> ...sales, production, and research and development.

Many writers in journalism and literature drop the last comma in a series, as long as no ambiguity results. Because of the complexity of items in scientific lists, though, there often are ambiguities. For that reason, I recommend leaving in the final comma. What about lists in which there wouldn't be ambiguities? I still recommend leaving in the comma for consistency's sake. In a document, a writer establishes certain punctuation patterns that after a few pages the readers come to expect. Leaving in the comma reinforces one of those patterns.

Although the rules for commas in a series and commas setting off contrasting elements are straightforward, many comma rules are hazy. For example, using a comma after an introductory phrase depends on the situation:

> In nine cases the people were infected by a rare strain of the virus that did not cause AIDS.

Placing a comma after "cases" is optional. Few readers would notice whether you did or didn't. Some introductory phrases, however, require a comma.

> When feeding a shark often mistakes undesirable food items for something it really desires.

You need a comma after "feeding."

When the comma rules are hazy, how do you decide whether to use commas? First, you should realize that the purpose of commas is to eliminate ambiguities. Comma rules aren't arbitrary. They have a specific purpose: to prevent readers from tripping over language. Therefore, when unsure about a comma, think about whether your readers would trip if the comma weren't there. Be consistent, though, in your use of commas. If you punctuate a sentence structure one way in the beginning of a paper, then try to punctuate that structure the same way throughout.

The Colon. Colons introduce lists:

> We studied five types of Marsupialia: opossums, bandicoots, koalas, wombats, and kangaroos.

Colons should not, though, break continuing statements.

> The five types of Marsupialia studied were: opossums, bandicoots, koalas, wombats, and kangaroos. *(incorrect)*
>
> The five types of Marsupialia studied were opossums, bandicoots, koalas, wombats, and kangaroos. *(correct)*

Besides introducing lists, colons are also used for definitions:

> The laboratory growth of this germanium crystal made possible a new astronomical tool: a gamma-ray detector with high-energy resolution.

The Semicolon. The semicolon is often misused in scientific writing. Some scientists toss semicolons into sentences whenever they're unsure what punctuation to use. The semicolon is an optional piece of punctuation—you don't have to use it. In fact, many good writers don't. The semicolon has two specific purposes, though. First, it connects two sentences closely linked in thought:

> There is no cure for Alzheimer's disease; it brings dementia and slow death to thousands of Americans every year.

Second, semicolons separate complex items in a list:

> Four sites were considered for the research facility: Livermore, California; Albuquerque, New Mexico; Los Alamos, New Mexico; and Amarillo, Texas.

Note that commas could not effectively separate these items, because each item contains a comma.

The Dash. The dash, or em-dash, sets off parenthetical remarks that cannot be set off by commas:

> The unique feature of the design is a continuous manifold, which follows a unidirectional—as opposed to serpentine—flow for the working fluid.

Dashes are also used to set off end phrases and clauses that would be ambiguous if set off by commas:

> After one year, we measured mirror reflectivity at 96%—a high percentage, but not as high as originally expected.

Be careful with the dash. Too many dashes will break the continuity of your writing. Also, note that dashes (—) are different from minus signs (–) and hyphens (-).

Quotation Marks. In the United States, end quotation marks go outside of periods and commas. This rule confuses many scientists and engineers because in Great Britain, quotation marks often appear inside of periods and commas.

U.S. Mistake:	According to Pauling, "Science is the search for truth".
U.S. Correction:	According to Pauling, "Science is the search for truth."

Hyphens. Compound words are common in scientific writing. Sometimes you can find the accepted spelling of these compounds in the dictionary. Many times, though, you can't. In such cases, you have to decide whether to hyphenate. Should you write fly ash or fly-ash? flow field or flow-field? cross section or cross-section? There are no clear-cut rules here. Many compounds start out as two words and then acquire hyphens after years of use. Although no clear-cut rules exist, identifying the compound's part of speech (noun or adjective) will help you decide.

The trend in spelling compound nouns is away from the use of hyphens because hyphens make the writing appear more complex. Therefore write

cross section
flow field
fly ash

When compounds appear as adjectives in front of nouns, the trend is to use a hyphen to avoid misleading the reader. Therefore, write

cross-sectional measurements
flow-field predictions
fly-ash modeling

References

Bernstein, Theodore M., *The Careful Writer: A Modern Guide to English Usage* (New York: Atheneum, 1965).

Fowler, H. W., *A Dictionary of Modern English Usage*, 2nd ed. (Oxford: Oxford University Press, 1965).

Funk, Robert, Elizabeth McMahan, and Susan Day, *The Elements of Grammar for Writers* (New York, Macmillan Publishing Co., 1991).

Hodges, John C., W. B. Horner, S. S. Webb, and R. K. Miller, *Harbrace College Handbook*, 12th edition (Fort Worth, TX: Harcourt Brace, 1994).

Sabin, William A., *The Gregg Reference Manual* (New York: McGraw-Hill, 1985).

Appendix B

A Usage Guide for Scientists and Engineers

"Usage" refers to the selection of the proper word. Some selections are clear-cut. For instance, no matter what reference book you consult, "its" is defined as "of it" and "it's" is defined as "it is." Other choices are not so clear. For example, some liberal sources, such as *Webster's New Collegiate Dictionary*, treat "farther" and "further" as synonyms, while conservative sources, such as *Elements of Style*, do not. Because scientific writing is more traditional than other kinds of writing, you should probably lean to the conservative side in your usage. This appendix presents a guide to common usage questions. At the end is a list of traditional usage references.

-ability words: A word ending in "ability" is a signal that you can tighten the sentence by having the word "can" precede the verb buried in the "ability" word. Although examples such as "capability" are not offensive, pretentious constructions such as "operationability" and "manufacturability" are. Rewrite the sentence using "can operate" and "can manufacture."

abstract nouns: nouns that offer none of the five senses to the audience. Common abstract nouns in scientific writing are "environment," "factor," and "nature." If you have to use an abstract noun, ground it with an example.

active voice/passive voice: In general, the active voice (having the subject perform the action) is a more natural way to communicate because it is crisper and more efficient than the passive voice. However, for those occasions in which the subject of your writing is acted upon, the passive voice is more natural.

affect/effect: "Affect" is a verb and means to influence (note that in psychology, "affect" has a special meaning as a noun). "Effect" is usually a noun and means a result. Occasionally people use "effect" as a verb meaning to bring about: "He effected the change of orders."

alright: not accepted usage. Use "all right."

always: a frightening word because it invites readers to think of exceptions. If an exception exists, your readers will find it, and your authority will be undercut. You should go in fear of absolutes.

approximately: appropriate when used to modify a measurement's accuracy to within a fraction, but inappropriate when applied to a situation such as "approximately twelve people." Does the writer mean 11.75 or 12.25 people? In such cases, use the simple word "about."

center around: The phrase "center around" makes no physical sense. You should either use the phrase "center on" or "revolve around."

clichés: descriptive phrases that have become trite. Common examples include "come up to speed" and "sticks out like a sore thumb." If a descriptive phrase sounds cute, avoid it.

component: often can be replaced by the simple word "part."

comprise/compose: "Comprise" literally means to include. Most conservative sources such as *Elements of Style* hold to that literal definition. For that reason, conservative sources insist on the whole comprising the parts, not the reverse. Likewise, conservative sources shun the phrase "is comprised of."

conjunctions: Conjunctions, such as "and" and "but," are powerful words that connect words, phrases, and clauses. Is it proper to begin sentences with conjunctions? Although some formal scientific journals frown on this usage, many respected publications, including *The New York Times,* allow it. With time, this usage will likely become accepted everywhere.

continuous/continual: "Continual" means repeatedly: "For two weeks, the sperm whales continually dived to great depths in search of food." The word "continuous" means without interruption: "The spectrum of light is continuous."

criterion/criteria: "Criterion" is the singular form, and "criteria" is the plural form. Note that this word comes from Greek, which explains the unusual plural form.

data: a plural form of "datum," a Latin word. Because "datum" is no longer used in English, many sources consider "data" acceptable as either singular or plural. Many conservatives refuse to budge on this word. If you need a singular form and do not want to ruffle feathers, spend a few extra words and write "a data point."

facilitate: a bureaucrat's word. Opt for the simpler wording "cause" or "bring about."

farther/further: Conservative sources distinguish between "farther" and "further," advocating that "farther" be used to indicate distance, and "further" be used for all other variables: time, intensity, depth of meaning. However, many conservative sources, including Bernstein [1965], admit that "further" will eventually become accepted for all uses.

fewer/less: In general, use "fewer" for items that can be counted and "less" for items that cannot. For that reason, write "fewer cells," "fewer errors," and "fewer fish in the stream." Likewise, write "less water," "less air," and "less foliage." Exceptions include sums of money and time: "less than five years ago" and "less than 1 million dollars."

first person: the use of "I" or "we." Occasional use of the first person can help reduce the unnatural use of the passive voice. As long as the emphasis remains on the subject of the writing, there is nothing inherently wrong with using the first person. You should understand, though, that some editors and managers (to their deaths) will forbid its use in scientific writing.

however: an adverb that has the same meaning, but not the same sentence function, as the coordinating conjunction "but." A coordinating conjunction can join two independent clauses; an adverb cannot. Is it proper to begin a sentence with "however"? Yes. As a transition word at the beginning of a sentence, "however" is accepted by all but the most conservative of readers.

implement: a pet verb of bureaucrats. Consider substituting "put into effect" or "carry out." These verb phrases are old and simple. They're the verb phrases that Winston Churchill would have used.

interface: the interstitial boundary between two systems, planes, or phases ("a computer interface" or "an oil-water interface"). Not acceptable is to use "interface" as a verb meaning to meet. The idea of two people or, worse yet, a group of people interfacing is unprofessional.

irregardless: not accepted usage. Use "regardless."

its/it's: "Its" is the possessive form of the pronoun "it" and means "of it." "It's" is a contraction and means "it is." A few hundred years ago, someone decided on these meanings. Accept them; learn them; write them.

-ization nouns: often pretentious. You should challenge "-ization" nouns and search for simpler substitutes. For example, replace "utilization" with "use." In cases where you have monstrosities such as "prioritorization," you should rewrite the entire sentence in forthright English.

-ize verbs: as with "-ization" nouns, often pretentious. Although some verbs such as "optimize" are forthright, other verbs such as "prioritorize" and "utilize" are pretentious. Opt instead for short, old words such as "rank" and "use."

-ized adjectives: as with "-ization" nouns, often pretentious. You should challenge "-ized" adjectives and search for simpler substitutes. For example, replace "discretized" with "discrete" and "individualized" with "individual."

like/as: "Like" is a preposition and introduces prepositional phrases: "Like Earth, Mars has an elliptical orbit." The word "as" is a conjunction and introduces clauses: "In Bohr's theory, the electron orbits the nucleus as a planet orbits a star."

never: a frightening word because it invites readers to think of exceptions. If an exception exists, your readers will find it, and your authority will be undercut. You should go in fear of absolutes.

only: a tricky word (sometimes adjective and other times adverb) that changes the meaning of a sentence by its position in the sentence. Check the position of "only" in the sentence to make sure that it modifies what you want it to modify.

phenomenon/phenomena: "Phenomenon" is the singular form, and "phenomena" is the plural form.

possessives: For single nouns, form the possessive by adding an 's: "a person's fingerprint," "someone else's decision," "your boss's authority." Exceptions include a handful of old people and places in which the possessive form does not add an *s* sound to the pronunciation (for instance, "Mount St. Helens' eruption" and "Archimedes' Principle").

principal/principle: "Principal" can be either a noun or an adjective. As an adjective, "principal" means main or most important. "Principle" appears only as a noun and means a law, as in "Archimedes' Principle."

redundancy: needless repetitions of words within a sentence. Common redundancies in scientific writing include

(already) existing	empty (void)
(alternative) choices	first (began)
at (the) present (time)	introduced (a new)
(basic) fundamentals	mix (together)
(completely) eliminate	never (before)
(continue to) remain	(none) at all
(currently) underway	(still) persists

stratum/strata: "Stratum" is the singular form, and "strata" is the plural form.

unique: an absolute that does not need a preceding modifier. Either something is unique or it is not. For that reason, the phrases "very unique" and "somewhat unique" make no sense.

weak verb phrases: In general, the smaller the verb phrase, the stronger the verb phrase. For that reason,

is beginning	→	begins
is used to detect	→	detects
performed the development of	→	developed
made the decision	→	decided
made the measurement of	→	measured

which/that: Use "that" for defining clauses: "We will select the option that has the highest thermal efficiency" (the clause tells which one). Use "which" for nondefining clauses: "We will select Option A, which has the highest thermal efficiency" (the clause adds a fact about the known option). Note that you separate "which" clauses from the rest of the sentence with commas.

writing zero: a phrase that adds nothing to the sentence and can be cut without loss of meaning. Common examples include "the fact that" and "the presence of."

References

Bernstein, Theodore M., *The Careful Writer: A Modern Guide to English Usage* (New York: Atheneum, 1965).

Fowler, H. W., *A Dictionary of Modern English Usage*, 2nd ed. (Oxford: Oxford University Press, 1965).

Sabin, William A., *The Gregg Reference Manual*, 6th edition (New York: McGraw-Hill, 1985).

Strunk, Jr., William, and E. B. White, *Elements of Style*, 3rd edition (New York: Macmillan, 1979).

Glossary

active voice: a verb form in which the subject performs the action in the sentence: "Jackals *attacked* a wildebeest calf."

adjective: a word that modifies a noun or pronoun: "Scientists were surprised when the *huge* earthquake and *unexpected* landslide triggered such a *massive* eruption."

adverb: a word that modifies a verb, adjective, or another adverb: "When Mount Pelée erupted, the lava gushed *so rapidly* through Saint-Pierre that *only* two citizens survived."

appositive: a noun or noun phrase placed next to a noun to define or explain that noun: "*A condition resembling intoxication,* nitrogen narcosis begins to affect scuba divers at depths of 30 meters."

back matter: the sections of a formal report following the text. Examples are the appendices and the glossary. Everything in the back matter is keyed somewhere in the text. Other parts of a formal report are the front matter and main text.

connotation: the implied or associated meanings of a word.

denotation: the dictionary meaning of a word.

dependent clause: a clause that cannot stand alone as a sentence. These clauses begin with words such as "although," "when," "because," and "if": "*Because anemometers are so delicate,* they cannot be used to measure directly the wind speeds of tornadoes."

format: the way in which a document or presentation is arranged. Format includes such things as the choice of typeface, the spacing between sections, the referencing system for sources. In engineering, there is no single ordained format. Whatever company you work for, whatever journal you submit to, you must arrange your document or presentation to meet that company's or journal's format.

front matter: the sections of a formal report preceding the main text's "Introduction." Examples of sections in the front matter are the title page, the table of contents, and the informative summary.

gerundial phrase: a participial phrase that functions as a noun: "In scuba diving, *ascending too quickly* will not allow your body enough time to dispose of the nitrogen that it has absorbed."

illustration: the meshing of words and images in a document. Illustration includes not only the presence of figures and tables, but also the captioning of those figures and tables as well as the introduction of those figures and tables in the text.

imperative mood: a command form of a verb in which the subject is an understood "you." An example would be as follows: "*Turn* on the computer." The subject of this sentence is "you," even though the word "you" isn't explicitly in the sentence.

infinitive phrase: a verb phrase in which the verb is coupled with the word "to": "*To calculate the energy*, you multiply the frequency by Planck's constant."

introductory series: a list of nouns at the beginning of a sentence that defines the subject of the sentence. The introductory series is separated from the rest of the sentence by a dash: "Nitrogen narcosis, decompression sickness, and arterial gas embolism—these are the greatest dangers facing scuba divers."

jargon: a word, abbreviation, or slang that is particular to a company, laboratory, or group. Within the company, laboratory, or group, the expressions can be an efficient shorthand for communicating information. However, to outside readers, the expressions are often alienating.

language: the way we use words in writing and speaking. Language is more than just vocabulary; it includes the order of words, the lengths of sentences, and the use of examples.

main text: the sections of a formal report from the "Introduction" through the "Conclusions." In general, the main text should stand alone as a document written to the primary audience of the formal report.

mechanics: grammar, punctuation, and spelling.

noun: a word that identifies an action, person, place, quality, or thing. Examples include "flight," "scientist," "laboratory," "curiosity," and "oscilloscope."

participial phrase: a verb phrase in which the verb often ends in "-ing" (present) or "-ed" (past): "*Hunting at night*, tigers are rarely seen *making a kill*."

passive voice: a verb form in which the subject does not act, but is acted upon: "The wildebeest calf *was attacked* by jackals." Note that the emphasis of the paper from which this example arises is on wildebeests, not jackals.

point: a unit of measure for the size of type (1 point ≈ 1/72 of an inch). The text of most documents is set in type sizes between 10 and 12 points.

pronoun: a word such as "he," "she," or "it" that may be used instead of a noun.

sans serif: a typestyle, such as Helvetica or Optima, in which no short strokes stem from the upper and lower ends of letters. Sans serif fonts are often used in posters, visuals, and illustrations.

scope: the boundaries of a document or presentation. Scope includes what the document or presentation will cover. The limitations are those aspects of the topic that will not be covered.

serif: a typestyle, such as Times or Schoolbook, in which short strokes stem from the upper and lower ends of letters. This book has been typeset in a serif font called Palatino.

structure: the strategy of your writing. Structure includes the organization of details, the transition between details, the depth of details, and the emphasis of details.

style: the way you present information in your writing. Style includes such things as the way you organize details, the words you select, and the illustrations you choose.

syntax: the ordering of words within a sentence.

tone: whatever in the writing indicates the attitude of the writer toward the subject.

verb: a word that indicates action or a state of being in a sentence: "The shock *shattered* the volcano's summit and *was* responsible for the collapse of the mountain's north side."

Index

D

E

F

Q

R

S

T

Made in the USA
Lexington, KY
27 August 2015